# Practical Balancing of Rotating Machinery

# Practical Balancing of Rotating Machinery

**Derek Norfield**

2006

ELSEVIER

AMSTERDAM • BOSTON • HEIDELBERG • LONDON • NEW YORK • OXFORD
PARIS • SAN DIEGO • SINGAPORE • SYDNEY • TOKYO

Elsevier
The Boulevard, Langford Lane, Kidlington, Oxford, OX5 1GB, UK
Radarweg 29, PO Box 211, 1000 AE Amsterdam, The Netherlands

First edition 2006

ISBN 10: 1-85-617465-4

**British Library Cataloguing in Publication Data**
Norfield, Derek
  Practical balancing of rotating machinery
  1. Balancing of machinery    2. Machinery - Vibration
  I. Title
  621.8'11

**Library of Congress Cataloging-in-Publication Data**
A catalog record for this book is available from the Library of Congress

Typeset by Charon Tec Ltd, Chennai, India
www.charontec.com
Transferred to Digital Printing in 2009

# CONTENTS

## ABOUT THE AUTHOR

Since the mid 1960s, Derek has been involved in many aspects of vibration control and balancing including machine design, development, and sales. During his apprenticeship program at an electric motor manufacturer in England, he was introduced to balancing and worked on maintaining and updating balancing machines of an in-house design. In 1970, he became involved with the UK division of Reutlinger balancing machines from Germany. Vibration measuring, analysis and monitoring were a major part of those activities. In 1978, Derek moved to the USA to run Reutlinger's American operation. From 1984 to 2001, Derek was with the American Hofmann Corporation where he worked on application and product development over many industries. Currently he has a successful consulting and training business, helping customers improve their performance through balancing and other methods of vibration reduction and control.

## PREFACE

Do you need this book?

If you deal with dynamic balancing problems of any kind the answer is, yes!

When balancing does not solve a vibration problem there is a strong and urgent need to find the real cause.

When balancing is a bottleneck in production you need to find out how to make it more efficient.

When customers complain or you get warranty failures you need to look at tolerances and procedures.

This book is a practical introduction to vibration as it affects rotating equipment. We use a lot of pictures to help illustrate the teaching and minimize the use of mathematics and formulas. The aim is to give you, the reader, an understanding of the process and the solution to a problem. We do not teach the details of setting up a balancing machine since each manufacturer already provides that for their particular piece of equipment. We do, however, teach what the balancing machine is doing and why it gives the results you see. We cover in detail the causes of typical setup and operational problems to enable you to understand what needs to be fixed.

Many rotors that need to be balanced have to be mounted on some type of tooling which, in turn, is then mounted in the machine. This book covers what is needed to ensure your tooling is working properly, and whether it is able to accomplish the real world balancing tolerance you need to achieve. When balanced components are assembled, the fitting tolerances and assembly methods can cause errors many times greater than the balancing tolerance. This book will teach you how to minimize or eliminate the error build-up.

This is your answer book for all balancing questions. If you have a question the book does not answer, then you can email us and we will do our best to give you the information you need.

*Derek Norfield*

# 1

# Introduction

In the beginning . . .

There were no machines, life was simple; but work was hard!

Given a basic premise that humanity was created with the unique desire and ability to change the world, machines and processes were needed. People's desire to make work easier brought forth ideas, converted those ideas into hardware and the software (or instructions) that made the device useful. For every device there has to be instruction on how to use it effectively. Hardware and software have always been equally important, but not always equally respected.

The early development of machines is not well documented and perhaps took the following course to bring us to present day requirements.

 'Life was simple; but work was hard!'

## The first machines a brief history

The first manufactured machine may have been a flint knife. In this case a non-manufactured machine (the knapping stone) was a prerequisite and the instructions for duplicating the knife were also vital. The learning curve would be long and teaching would have been a 'one-on-one' situation. With knives or spears in hand the first engineers had to develop programming with a 'Graphic User Interface (GUI)' to coordinate a group operation and the first cave wall pictures of how to hunt soon followed.

Such technological development led to continuous improvements in everybody's lifestyle. The pace of development really got moving once a critical stage was reached – the chair was invented.

You may not have considered how important this development would be, but once a person got comfortably seated their active brain could really get to work. Now in a comfortable chair he or she soon realized the urgent need for packaged food, drink and entertainment. This led directly to the beverage industry, the snack food industry and ball games.

As the number of tribes involved in ball games increased, there was the need to make the chair easy to move and this in turn led to the development of the automobile. Naturally this in turn led to the development of the entire petrochemical industry.

Two more technologies were urgently needed. Containers for the snack food, and news of remote ball games and personal hygiene improvements led to the development of the paper industry. Complaints from the females about the growing piles of debris around the men's chairs led to the development of vacuum cleaners.

As you can clearly see, the chair led to the rapid development of all of modern industry. Common to all of these industries was a need for rotating equipment for pumps, fans, rolls and wheels. In order for these machines to run smoothly we had finally to develop the balancing machine.

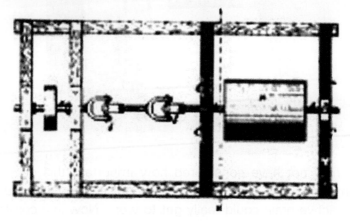

Credit: Henry Martinson's balancing machine. Patented in 1870.

## The need for balancing

Until the 1870s, balancing was not known, or needed, in almost any industry. The main exception is the waterwheel and windmill. These wheels had to be statically balanced so that they rotated at a constant speed, especially when the water or air flow was low. Unbalance in the sails and wheels caused fluctuations in the grinding wheel speed and the resulting meal flour would have an uneven texture. An out of balance waterwheel might not even turn after being stopped for a while and water had drained from the pockets. There has to be enough water mass to overcome both friction and the wheel unbalance.

Here is the first important principle about balancing – it is NOT an expensive add-on operation! Balancing is a cost saving, efficiency boosting process.

The first miller who learned how to balance the sail or water-wheel achieved operation in lower wind or water flow conditions, and had more consistent grinding conditions – more 'uptime' as well as a higher quality, higher value product.

Hardware needed: block of wood, nails and hammer. Software needed: knowledge of how much wood to add and where to add it.

Then, as now, the itinerant balancing consultant was told that the prices were too high and that their services were not needed. However, the success of an enlightened customer led to market dominance and eventual takeover of any unbalanced rivals.

In order to protect his or her knowledge, the balancing consultant would add rituals (normally called procedures) and misdirection (which covered any mistakes) so that the simple process of balancing became regarded as a black art and the consultant was renamed 'the balancing wizard'. As you read this book you will have the opportunity to learn the simple rules that will enable you to astound your peers with machinery that runs smoothly and reliably – but like the wizards of old, you too will find that to most people your instructions will sound like spells and incantations and the results will be like magic.

From the dawn of time until the middle of the nineteenth century, technology was based on the power of water, wind, oxen or horses. But that was all to change as a new breed of wizards called 'engineers' started to harness unheard of powers.

## The industrial revolution

Three technologies were about to emerge to change the world.

The advent of the steam engine brought far reaching changes to society, and to the field of balancing. Early steam engines ran at speeds from 10 to 50 rpm. As steam engines ran faster, the requirements for balancing the flywheels increased. By 1869 when the US Transcontinental Railroad was completed across the USA, locomotives could reach the incredible speed of 30 mph. In 1884 Charles Parsons developed a 10 hp steam turbine that ran at 18,000 rpm. Due to the lack of balancing technology he made the bearings elastic so the turbine would actually spin about its axis of mass.

Parsons Steam Engine, 1884.

While this 'revolution' in steam power was happening, Michael Faraday invented the DC electric motor (around 1850).

First Faraday motor.

Then in 1887, Nikola Tesla invented the first AC motor. The electric motor raised the technology of balancing since speeds went to 900, 1800, the 3600 rpm and higher.

First practical AC Motor, 1887.

The electric motor was used in simple machines, whereas the steam turbine was a large complex system. By 1930 the steam turbine was basically used for electric power generation and marine propulsion, and the electric motor was the prime mover for the vast majority of applications. These motors were installed on machinery that was now running faster than ever before and balancing became of vital importance.

The third technology is, of course, the internal combustion engine. Steam power uses combustion external to the 'engine' but the Diesel and Otto engines moved the site of combustion to the inside. In 1897 the first working diesel engine was tested with an output of 20 hp and an efficiency of 26 per cent (more than double that of the current steam engines).

Daimler engine installed in vehicle, 1897.

In 1876 Dr. Otto developed the six stage, 4-stroke cycle of combustion used in the Wright brothers' flyer and the ubiquitous
gasoline engines that drive our cars today. In 1885 Otto's
student, Daimler, developed the engine to run at 900 rpm and
put it in one of the first automobiles.

In 50 years of industrial revolution rotational speeds went from
tens of revolutions per minute to thousands of rpm. From a
waterwheel at 5 rpm to a steam engine at 24 rpm then a diesel
engine at 250 rpm, a gasoline engine at 900 rpm, an electric
motor at 1800 rpm to a steam turbine at 18,000 rpm.

One final technology was to come: Internal combustion was
brought to the turbine.

USAF Stealth Fighter F117, Twenty-first Century.

The gas turbine was the final enabling technology for efficient electric power generation in transportable systems and for high speed aviation – a technology with speeds up to 100,000 rpm. Just 100 years after the first engines we have supersonic jet aircraft and standalone turbine generating systems the size of a refrigerator that just need a fuel supply to output 100 kw of electrical power.

None of these systems we take for granted today would be practical without dynamic balancing technology. Balancing is one of the 'enabling' technologies that allow development and production of more powerful and efficient equipment.

---

**Consider this**

If a person from ancient Egypt could be brought forward in time they would have understood all the technology in normal use all the way from 2000 BC to 1900 AD. In the last 100 years we have gone from oil lamps, semaphore flags, sailing ships and horse transport to high intensity lighting, worldwide wireless telephone communication, nuclear submarines, and 'planes, trains and automobiles.

---

Sixty kilowatt refrigerator sized generator. Courtesy of Capstone Turbine Inc.

## The history of balancing

The technology of balancing has paralleled the development of prime movers. When the fastest machine was a waterwheel no equipment was needed. Gravity was available and the technique was to add mass to whatever part of the rotor went to the top. If the rotor was turned 90 degrees and did not move when released then it was balanced. If it moved then more mass was added.

Waterpower from gravity.

As late as 1920 most balancing was trial and error. Balancing machines using mechanical resonance or amplification were operational but had to be meticulously calibrated to give useful information. Operation of such a machine was a mixture of learned skill and intuition. An example of typical technology would be paper machinery rolls that were balanced by holding a piece of chalk close to the rotor, where it first touched was the high spot. As a rotor was progressively balanced the length of the chalk mark on the circumference increased. When the result was a continuous line the roll was in balance.

Electrical technology brought phase generators and compensators that enabled the mechanical vibration due to unbalance to be compensated to 'zero'. The compensation amount and phase was the exact inverse of the unbalance. When technology moved from electrical controls to electronic controls, the machines became much more accurate, easier to use, and

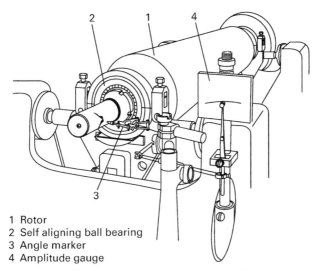

1 Rotor
2 Self aligning ball bearing
3 Angle marker
4 Amplitude gauge

Early mechanical balancer. Chalk in marker touches highest point, dial indictor shows highest amplitude. Operator skill level needed – high!.

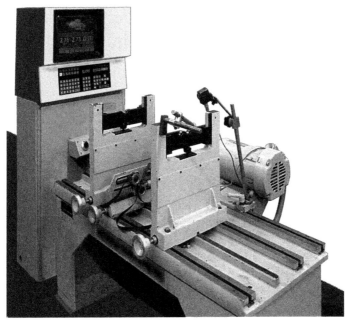

Typical modern hard bearing, belt drive, balancing machine with microprocessor based instrumentation.

more reliable. Finally, the computer completed the transformation of the system. Although the control system gets the glory a well designed and precise mechanical structure and transducer system is even more important to accurate and repeatable balancing.

To those of you who laughed at the 'wizard' illustration – can you explain what goes on in the modern computerized balancing instrumentation? The science and science fiction author Arthur C. Clarke (do you remember '2001' and the computer named HAL) summed it up with this statement: 'Any sufficiently advanced technology is indistinguishable from magic'.

# Any sufficiently

# Advanced Technology

# is Indistinguishable

# From Magic.

As a child growing up in England in the 1950s, technology around me seemed incredible. We had television (black and white), telephones (operator connect), heaters in cars (optional), smokeless coal for heating, and ballpoint pens that only leaked half the time (compared to fountain pens that leaked all the time). I was born the same year that the transistor was invented

(1947) but didn't care about that until the advent of the transistor radio. I still remember my first one, it was an ugly green color, ran for about 2 hours on a battery, had trouble locking onto a station and sounded really tinny – but it was the ultimate in current entertainment technology.

High Tech, circa 1950.

While we marveled at this device, Sputnik flew overhead – incredible for 1957 technology. As an apprentice in the 1960s I worked on my first balancing machine – built in-house, soft bearing and with tube type electronics. It would sometimes run an entire shift without breaking down. You could be proud if you could set up for a new job in less than a half hour. Ten minutes of every hour was allocated to calibration checking to be sure it had not drifted (an unknown problem today). Back then, the same as today, a speck of dirt on a support bearing could mess up the results more than any electronic problem. The more things change, the more they stay the same.

The twenty-first century has brought high technology comput-
ing, improved mechanical design and even active continuous
balancing to the market but has not changed the underlying
tolerance requirements based on rotational speed and bearing
loads. The other thing that has not changed is the requirement
that the mechanical systems of the machine also work flaw-
lessly (more on that later).

## Why do we balance?

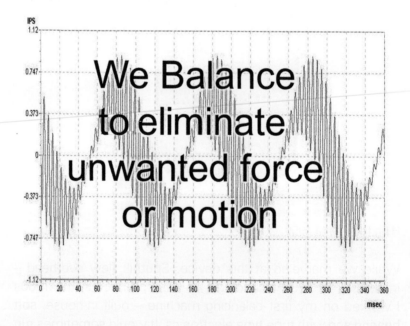

Balancing is a way to reduce vibration and bearing loads to
improve performance and reliability. There are three basic
benefits from balancing to the proper balancing tolerances.

Balancing reduces the loads on the bearings. Bearing life is
proportional to load and speed. By limiting the centrifugal

forces to less than 10 per cent of the static load bearing life is maximized.

Balancing reduces vibration. Vibration causes parts to become loose, generates noise, produces a perception of low quality and, in the case of products such as power hand tools, may even be harmful to health.

Balancing allows performance to be improved by increasing operating speeds. The higher speeds are possible due to the reduced bearing loads and elimination of internal bending stresses on longer rotors.

When we look at products such a computer disk drives, electric drills, aircraft engines and even the cooling fans on our car engines we soon see that none of these products would perform as well, or even be practical, without proper balancing.

The argument is often heard that 'Balancing is an additional operation', 'it will cost too much to implement balancing of our parts' but this thinking is incorrect. Balancing is still seen as a form of black magic, something to be avoided. People some-how connect the balancing specialist with Shakespeare's Weird Sisters – in his play 'MacBeth' – and their bubbling caul-dron. Be assured we do not deal in potions and powders but in graphs, spreadsheets, and cost effectiveness.

## Balancing is not an additional expense!

Balancing is often seen as a cost factor when, in fact, it is a cost reduction process. Often we hear companies complain at the cost of balancing. It is not seen as a part of essential manufacturing. For example, an electric motor rotor needs a shaft, it needs the laminations and electrical components, and it needs bearings. All that balancing does is put a few holes in or add some weights to the product.

If we could manufacture perfect rotors we would not need to balance – what would be the cost of manufacturing perfect rotors?

What would be the cost, and availability, of perfect material? Laminations for the above-mentioned rotor have tolerances on concentricity and thickness. Buying laminations to tighter specifications increases the cost, reduces the number of suppliers, jeopardizes reliable delivery and increases inspection requirements.

By definition balancing is the correction of manufacturing problems. No manufacturing material or process is perfect and errors in manufacturing and assembly combine and build up with the result that the final product has too much noise, vibration, shaft bending, bearing load or performance loss to meet the final testing requirements. Balancing is an enabling technology and not an additional cost.

   'Balancing is an enabling technology'.

Balancing enables performance standards to be maintained with coarser manufacturing processes than would otherwise have been needed. In many cases it is simply not possible to achieve the desired results without balancing.

An engineer looking for the most economic way to manufacture a product has to examine the cost/benefit of each process and operation. In the first assessment balancing might not be included, and only a lack of acceptable product performance forces a balancing operation to be added. This is considered an added cost – WRONG! The correct situation is that the original process was not capable of producing an acceptable part.

Solutions such as changing material, tightening machining tolerances or changing the manufacturing process were more costly than adding the balancing operation. Balancing is added to a manufacturing process because it is the lowest cost method for achieving the required performance.

When vibration is still a problem then ignorance produces a knee jerk reaction 'cut the tolerance'. Sometimes this can be beneficial but usually it increases the cost without producing any benefit. Often a balancing operation is done at a stage of manufacture when it is convenient to do it rather than when it is most effective. For example a fan blade assembly may be balanced and then installed on a shaft and secured with a bolt from the side. The bolt pulls the fan to one side and twists it so it has a 'wobble'. When the fan and shaft is run up to speed it vibrates. First (wrong) solution – tighten the balance tolerance. Second solution – balance the assembly. Third (optimum) solution – change the clamping. Often balancing problems are actually design problems in disguise.

'Balancing problems are actually design problems in disguise.'

A number of years ago we were working with a manufacturer of computer disk drives. After insisting that balancing was not required, they needed to increase the disk capacity which meant closer track spacing. Vibration limited the tracking and balancing was added. The immediate benefit was perfect operation at the new higher capacity. This led to an additional narrowing of the track spacing followed by an increase in rotational speed which reduced access time and increased the maximum data rate. Analysis of the product and the company two years after the addition of balancing showed that without balancing the company would have ceased to exist (left behind in the race for performance). Conversely, the early adoption of balancing

led to the company growing more rapidly than any of the com-
petitors. Since this involved the sale of about 200 balancing
machines it also made us very happy. What made this even
more interesting was the development of the balancing
machine, correction system, and balancing procedures that
made its price and performance accuracy and cycle time,
advance to the point that it was profitable both for us and for
the customer.

Caption Standard hard disk drive showing disk and heads.

How far has disk drive technology come in 20 years? We
started with the original IBM PC AT with a 10 MB drive. Today a
leading edge office PC may have up to a 500 GB hard drive (that
is a 50,000 times greater capacity) at the same price as the
original but with much improved reliability and incredibly fast
data transfer rates. And this was only possible by the addition
of high precision balancing.

In fact, the IBM AT drive was already a huge step from the large (12″ diameter) disk packs and the 'Winchester' disk drives of 10″, 8″ and 6″ that preceded the 5.25″ diameter format. When those original disk packs were introduced they could only hold a few megabytes and 30 years later we have 1″ diameter drives holding 2 gigabytes that fit inside a digital camera. And these drives are being replaced by solid state memory.

Now in case you were thinking that the use of solid state memory eliminates balancing – think again. In order to make these chips they are laser (or electron beam) etched in a vacuum. To maintain the fine line widths of less than 0.5 micron the vacuum pumps have to be balanced to incredibly small

Turbomolecular pump, runs at 80,000 rpm and removes the last 10 per cent of atmosphere from the chip engraver systems.

tolerances. Not only are the tolerances tiny but the balancing is done at high speed and uses temperature compensation and prediction routines that ensure that the pump will still be in balance after it reaches operating temperature.

More balancing wizardry!

## Vibration, causes and effects

An unbalanced rotor will tend to vibrate while it is spinning. We measure the unbalance by measuring this vibration. However, not all vibration is caused by rotor unbalance. From this we must realize that to understand balancing we first need to have some knowledge of vibration.

**What is vibration?**
- A regularly reversing movement
- A measurable amplitude
- A measurable range of frequencies
- A measurable intensity
- A measurable acceleration

Vibration is a repetitive back and forth motion. It is either forced (driven by external input) or natural (resonant). Musical instruments produce loud sounds by resonant amplification of a small input. A diesel engine vibrates due to forces from combustion energy and piston movement. Naturally some vibrations are desirable while others are objectionable – and sometimes this is a matter of opinion!

One person's music may be another person's noise. One person's weed may be another's desirable plant. My wife is a master gardener but I subscribe to the PIU school of gardening: PIU – Pull It Up – If it grows again it was a weed.

Unwanted stuff always seems to grow back. Unwanted vibration always tends to come back. The four 'basic food groups' of vibration are bearings, misalignment, looseness and unbalance.

**Rolling element bearing**

We need to be able to identify the undesirable vibration. We have to learn about the vibration characteristics and to know how to pick the tool to use for the specific problem. It is easy to tell when a machine has excessive vibration, right, just put your hand on it!

Balance Weight

# Miralignment

Parallel                                    Angular

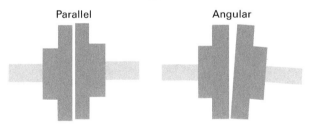

Now if the machine is a steam turbine you might have just burnt your hand! If there is a bearing that is dying you might not sense the high frequency but low amplitude vibration. Even if you can sense that the vibration is too high how do you diagnose the root cause?

   'You have to get rid of the root'

Picking the heads off dandelions does not kill them, the root goes too deep. To get rid of a dandelion or a vibration problem you have to get rid of the root! Vibration is a symptom and not the problem. We have to get to the root!

Get the root

A doctor will check your pulse, take your temperature, measure your breathing rate, listen to your lungs and check your system response (knee jerk reaction). Having a temperature is good – if it is the right temperature. Having a pulse is good (measured by vibration of vein) if it is at the right frequency. Breathing is good but a vibration signal superimposed on the normal vibration (sound is vibration) may indicate fluid contamination. Having a knee jerk response is good – if it is to the doctor's hammer hitting below our kneecap. Sometimes we have a knee jerk response to the wrong stimulus.

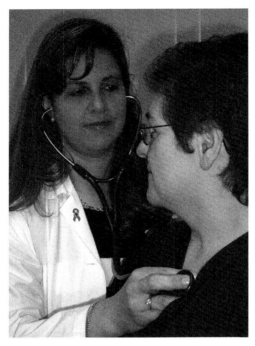

Vibration test equipment and 'expert system' for diagnosis of problems.

What we are talking about is diagnosis. The doctor looks, listens and tests in order to diagnose the root cause of the problem. You will have noticed that he uses his ears as a vibration analysis tool. The stethoscope is a mechanical amplifier of vibration signals. It is also part of the toolkit of a vibration specialist but we also use much more sophisticated tools. Where the doctor might use a CAT scan we may use a vibration analyzer. Temperature can be an indication of a bearing problem and any machine that is noisy is trying to tell us something.

Where do we start? What are the 'vital signs' we are looking for? The doctor starts with simple measurements and we will start with simple vibration.

## Simple vibration

A mass suspended on a spring is about the simplest vibration system. If energy is added (pull down the weight and release) the result will be a vibration at a frequency determined by the size of the mass and the strength of the spring.

In a theoretical physics class we may imagine a frictionless device however the spring functions due to the stress in the steel, and there is movement at the molecular level. We are not in a vacuum so there is air resistance to the movement. Each cycle of the spring mass vibration will dissipate some vibration and so the amplitude will gradually decrease – the frequency stays the same but lower energy leads to lower amplitude.

A grandfather clock uses the same system with the pendulum acting as the mass and the spring – gravity pulls in a constant direction and the curved motion about the fulcrum is always pulling the mass downwards. The clock keeps running

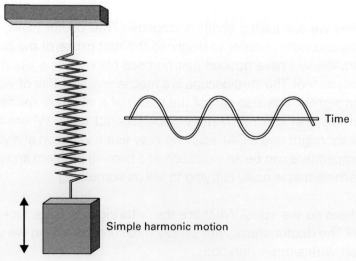

Simple harmonic motion

It is an example of Simple Harmonic motion (SHM).

due to the energy input from the escapement. To change the frequency we change the length of the pendulum. The simple spring mass and the pendulum have what is known as Simple Harmonic Motion (SHM). It is simple because there is just one single frequency involved. It is harmonic because it has smooth acceleration and deceleration.

In all vibration and balance situations we deal with mass not weight. *Mass* is a constant while *weight* is just the reaction due to gravity. We therefore have an object that may be called a 'weight' but it actually has a mass (gram, ounce etc.). If we take the spring and mass into space so we have a microgravity environment the weight is gone but the mass is unchanged.

Weight may be measured on a scale (say a spring balance). The weight will vary depending on location (sea level, in a mine, in a space station, on the moon etc.). The mass (inertia) is constant. Weight is a STATIC condition. Motion is a dynamic environment – try and weigh something while you walk around carrying the scale!

Simple vibration is specifically defined as being a sine wave and is comprised of a single frequency. When vibration contains more than one frequency it is complex.

Simple Harmonic Motion is a specific, very special type of vibration. If we want to generate SHM an easy way to do it is by using a crankshaft, connecting rod and piston – just like in a steam engine or even an automobile engine. If we take a railroad engine as our example we can see the operation (in our automobile engine it is hidden). The wheel of the engine rotates. It is a repetitive motion. The piston moves in a straight line from point A to point B and back to point A. Since it is connected to the wheel by the connecting rod and the rod does not change length then the position of the piston is

determined by the position of the wheel. There are a number
of conditions that we can measure or deduce from this
motion system.

When the piston is at the end of its travel it comes to a
smooth stop.

At the center of travel it is moving at maximum speed but it is
not accelerating or decelerating.

The direction of motion changes every 180 degrees of wheel
rotation.

Velocity is a measure of speed and direction so even though
the piston reaches the same speed four times per revolution
it reaches the same velocity twice per revolution.

The acceleration direction changes every 180 degrees of wheel rotation.

The acceleration is a maximum when the speed is a minimum.

The acceleration is a minimum when speed is a maximum.

The acceleration is a maximum when the wheel has moved 90 degrees from the position of maximum piston speed.

We can express the piston speed and direction in terms of the wheel position in degrees of rotation.

If the piston is at the ends of its travel (zero speed) at 90 degrees and 270 degrees then it will be at maximum speed at 0 degrees and 180 degrees.

There is a 90 degree phase difference between zero position and zero velocity.

The acceleration will be zero at 0 degrees and 180 degrees and therefore we can say that the acceleration has a 90 degree phase difference to the velocity and a 180 degree phase difference to the displacement (acceleration inwards increasing as displacement outwards increases).

There is a small difference in these motions due to the finite length of the connecting rod. A Longer rod will approach closer to the theoretical condition. If you wish to explore this in greater detail then please feel free to do so. It is not required! What is required is an understanding of the relationship between position of the piston (*displacement*), *velocity* of the piston and *acceleration* of the piston. The next graph shows these three aspects of motion with their phase relationship.

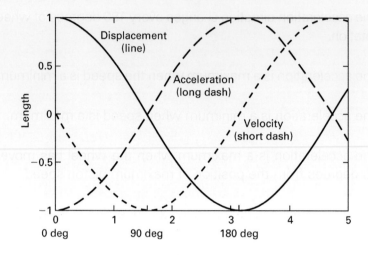

# Complex vibration

If we add a second spring and mass of different sizes to our system they will vibrate with different amplitudes and different frequencies. If we combine the two vibrations the result is a repetitive pattern but it is not a sine wave.

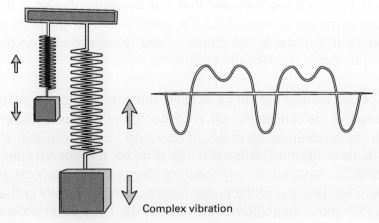

This is a complex vibration.

A major part of working with vibration is separating the vibration into its component parts. Modern vibration analysis equipment uses techniques formulated by an engineer named Jean Baptiste Fourier who was employed by the emperor Napoleon. As so often is the case, military research provides the impetus for technology breakthroughs. Fourier was working on heat conduction to improve Napoleon's cannons which were overheating. It was later found that the same analysis techniques worked for vibration. The Fourier transform essentially enables a complex vibration to be broken down into a series of sine waves, each with a specific frequency, amplitude and phase relationship. Those frequencies can be compared to rotational speeds and their multiples so we can determine the source of the vibration.

## Damping

A vibrating system will continue to vibrate unless energy is removed. The factor that removes vibration energy is damping. The effects of resonant vibration are much reduced by the addition of damping.

In the last 20 years huge advances in damping technology have improved the ride and handling of our automobiles. Some damping systems use active control of damper stiffness calculated from the output of a vibration sensor. In just milliseconds after a wheel hits a bump the shock absorber is stiffened to control the wheel motion and then relaxed to maintain a smooth ride.

Industrial systems also need damping to dissipate vibration energy. Damping usually involves materials that absorb energy through friction. A felt pad applied to a surface acts in this way by means of the densely packed fibers rubbing against each other.

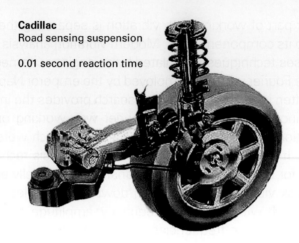

**Cadillac**
Road sensing suspension

0.01 second reaction time

Are there exceptions? One example of an extremely undamped system is a bell. Because of very low damping the bell rings for a long time and the energy is removed by interaction with the air.

If you grasp the edge of a ringing bell the sound stops almost instantly. That is the effect of adding damping to the system. In general an undamped system will tend to vibrate at a single frequency called the 'Natural' frequency, resonant frequency or critical frequency. The addition of damping reduces the vibration amplitude and also spreads the energy over a range of frequencies.

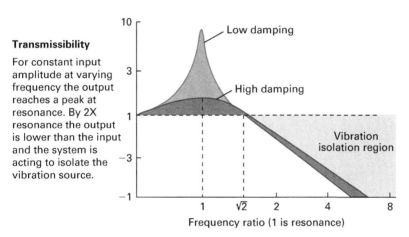

**Transmissibility**

For constant input amplitude at varying frequency the output reaches a peak at resonance. By 2X resonance the output is lower than the input and the system is acting to isolate the vibration source.

A vibration isolator allows a percentage of the source vibration to pass across it. To isolate the system from the source the isolator needs a resonant (or natural) frequency lower than the isolated frequencies. The isolator also needs a high level of damping to avoid excessive vibration at resonance. An automobile suspension uses soft springs and a high level of damping to provide a smooth ride.

# Why is vibration so bad?

Machinery is designed to perform a specific task to a given tolerance and vibration will normally have an adverse effect on the performance. A factory making microprocessor chips will use high speed turbo pumps to generate a vacuum. These chips have details narrower than a micron (one millionth of a meter) and vibration causes a high failure rate.

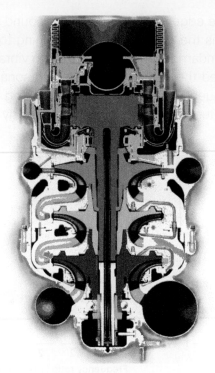

This turbopump runs at 40,000 rpm with small clearances. Vibration can cause a rub to occur and the result of that is catastrophic failure.

Turbine bearing life is limited by speed and load. Vibration forces can be many times the gravitational load on the bearings and cause early bearing failure. A grinding machine is employed to manufacture parts to a tight tolerance and fine surface finish. Vibration will cause ripples in the surface, greater wear on the grinding wheel and therefore a higher reject rate and part cost.

Applications such as a refinery, paper manufacturing, turbocharger production or airplane operation use vibration measurements to diagnose impending failures, shut systems down before catastrophic failure, improve productivity, and produce better quality products and services.

During operation of machinery and equipment parts wear, become loose, move and get a build up of contamination. Vibration tends to increase during operation. This vibration may be due to unbalance and unbalance may be caused by the actions involved in overhaul of the unit. Unbalance is not the only cause of vibration!

## What is the cause of vibration?

Many things can cause vibration – here are a few examples:

 'Balancing won't fix anything except unbalance.'

Misalignment, bent shaft, a shaft rubbing on the stator, damaged bearings, turbulence, cavitation, oil whirl, inadequate lubrication, loose mountings, worn gears, stator windings, broken rotor bars and unbalance.

Yes, sometimes a vibration problem is actually due to unbalance; but often vibration is due to other causes and balancing won't fix anything but unbalance.

Repeat aloud ten times –

'Balancing won't fix anything but unbalance!'

## Misalignment

If a motor is driving a pump and connected by a coupling there is an opportunity for parallel and angular misalignment. This will cause vibration at both 1× and 2× rotational speed in both radial and axial directions. Within the motor or pump a distorted housing can result in bearings that are out of line. Bearings that are not square on the shaft are a problem.

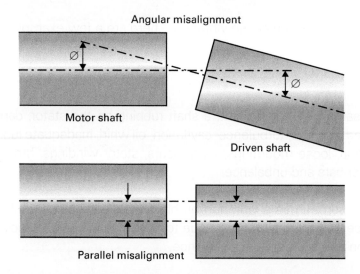

Angular misalignment

Motor shaft

Driven shaft

Parallel misalignment

## Damaged bearings

Installing a pulley with an undersize bore can damage the bearing, especially if the installation tool is a large hammer. The shock loads can pit the bearings or even crack a race or ball. I have seen bearings installed with a punch and hammer rather than a sleeve and the result is a bearing race that is distorted or not square on the shaft. The semi-technical term for this type of impact damage is the verb 'Brinell'. It implies that there may not be any visible damage but the material substructure has been changed and that wear will start to occur in a much shorter time.

What can we expect from a motor with improperly installed bearings?

> Noise, vibration, shortened life, rotor rub.

## Bent shaft

A shaft that is straight at room temperature may bend when running at full load, especially if there is an uneven heating effect. Excessive belt tension can bend a shaft (also it can cause rapid bearing failure). Longer shafts will sag if they are not kept turning slowly. Other shaft problems can result from machining errors. It is not uncommon for fork lift trucks to get off course and the resulting impact damage can be fatal to equipment.

## Turbulence

Ideally, a fan will have several blades that are equally spaced and identically shaped. In practice, each blade is slightly different. The result is that one section of the fan produces more thrust than another, culminating in a rotating thrust imbalance. A fan or pump that is mounted in an irregular housing will also have unbalance thrust or pressure.

Fan with bent blade, may be balanced
but it will vibrate at full load.

## Hydraulic (or aerodynamic) unbalance

An eccentrically mounted pump rotor will have variations in pressure at 1× rpm causing vibration that looks similar to unbalance. Look for smooth running dry but vibration when wet. These effects are generally termed aerodynamic or hydraulic unbalance. Today large or critical machines can be fitted with active continuous balancing systems that trim out the once per revolution forces. This feed back system will change the positions of a pair of eccentric masses in a housing and compensate the unbalance without stopping the rotor.

## Cavitation

Another problem with pumps is that a restricted inlet will generate a low pressure area at the impeller and cause the pump to cavitate. Cavitation erodes material and reduces performance. There are other causes of cavitation but they basically result from operating the pump away from its optimum flow

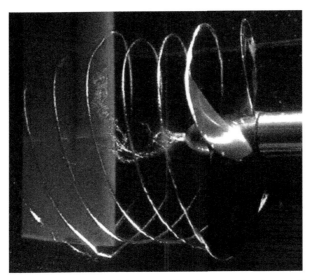

The bright circles are air bubbles resulting from cavitation on the tip of the rear face of the propeller. The lack of symmetry also shows that the blades are not of identical shape.

and head conditions. Cavitation produces a high frequency broadband noise not related to operating rpm. The pressure imbalance will also tend to cause seal wear and therefore shortened life.

## Oil whirl

When an oil film bearing has too much radial clearance and/or too small a radial load, the bearing may become unstable. The result is that the shaft will orbit in the bearing at just under 50 per cent of the shaft speed. The basic problem is that there is insufficient damping to dissipate the vibration energy.

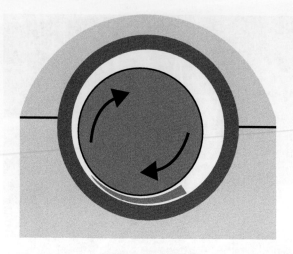

## Looseness

Machinery that is not bolted rigidly to its foundation will have excessive vibration. The foundation controls the energy absorption and a lack of coupling can result in very high amplitudes. Sometimes the vibration is at 2× or 3× shaft rpm but sometimes it is related to a resonance of part of the system and occur at other frequencies. The problem may be due to an irregular mounting surface or the use of a number of thin shims under a motor foot rather than a single piece of the correct thickness.

It is not recommended to go surfing near an oil slick.

## Bearing lubrication

You can have too much of a good thing. Too much lubrication causes bearing balls and rollers to skid rather than roll and this can generate excessive heat – over lubrication is a common problem. A lack of lubrication will cause noise and rapid wear.

Use of the correct lubricant (viscosity, grade) is critical. Many problems can be traced back to incorrect or contaminated lubricant.

# Worn or damaged gear

Gears have to be manufactured to very tight tolerances to operate quietly and without vibration. Slight errors from manufacturing or assembly can cause uneven loading and rapid wear. The wear patterns then cause vibration. If the gear ratios are not chosen carefully (prime number gears) the hunting tooth pattern where one tooth of the driver frequently hits one tooth of the driven gear will cause rapid wear and noise and vibration can result (e.g. with a 2:1 ratio tooth #1 on the driver will hit tooth #1 on the driven every second revolution).

# Unbalance

Unbalance causes radial loading on the bearings.

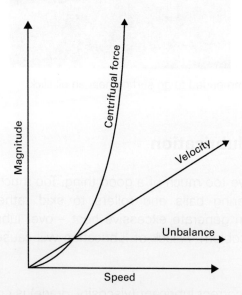

Unbalance is a property of the rotor and does not change with rpm. Vibration Velocity changes in direct ratio to the rotational speed. Centrifugal Force changes according to $(RPM)^2$.

How does a rotor become unbalanced?

Material builds up or erodes during operation. Foreign object damage and thermal effects, shaft sag, and rotor stator rubs may also occur.

Maintenance often means there is a change in operating conditions. The process of dismantling, replacing bearings, cleaning and reassembling will result in a different condition. This has the unintended result of further failure (Maintenance Induced Failure – MIF).

When there is no apparent reason for vibration or unbalance to have changed, it could be that excessive loads (torque or radial) have caused a crack to propagate through the shaft. This produces a change in stiffness at one radial location. If not remedied the final result is a broken shaft.

Unbalance is a property of the mass distribution of the shaft assembly (with bearings). If the unbalance has changed there must be some mechanical change that was the cause.

When you look at the circles for a few seconds you will see some unwanted motion.

# 2

# Eliminate unwanted motion!

## Introduction

When driving, or riding, in a car you want forward motion. That is the reason for being in the vehicle. You don't want noise, vibration, rattles, squeaks groans or screams. When operating machinery we usually want the rotational motion but not the side effects.

Scream now.

## How do we do that?

It is all a matter of tolerances, tooling, unbalance correction methods and locations, balancing machines, and many other details that affect the performance of a product.

Balancing may also be required due to the rotor design, or lack of design. Design errors, wide manufacturing tolerances, assembly procedures and rotor function, may cause large amounts of unbalance.

Generally, the addition of a balancing operation allows for opening up manufacturing tolerances. Sometimes this is carried too far and the result is a higher amount of unbalance than what can be efficiently or even practically removed.

The 'value added' by balancing comes from broadening of manufacturing tolerances and material specifications, elimination of warranty failures, increased product performance, and the customer's perception of a 'quality' product.

Up to now, we have been looking at the results of not balancing to optimum levels. Now we turn to the process of balancing.

## Balancing is Science, Technology and Art

Rotor balancing involves all three of the above parameters.

*Science* – theory and calculations
*Technology* – equipment and methods
*Art* – the secrets to effective use of the science and technology

Don't worry, its not that hard!

 "Its difficult to burn rubber when your wheels fall off!"

That theory and calculation stuff can be scary! It is not complex but it is essential to a real comprehension of what we are doing in the balancing process. We will cover this in enough detail to give you a good understanding of the reasons why we balance to certain standards.

Technology is the 'nuts and bolts' of what we are doing. An automobile would not operate without nuts and bolts. They are not always so exciting but it is hard to burn rubber when the wheels fall off.

When you know how to calculate the appropriate tolerance and can give the reason why – you are an 'expert'. When you have finished this book you will be an 'expert'.

The 'art' of balancing is where the need for creativity comes in. It is like painting a room. You measure very carefully, deduct areas of windows and doors and end up with the need to paint 300 square feet. You go to the paint store, pick up a gallon of paint and see on the labels 'coverage 200–400 square feet' – do you buy one can or two? (This, of course, depends on price of paint and how far away the paint store is. You also have to consider the probability of running out of paint just after the store closes).

In balancing we have to consider the cost of a tight tolerance or of a specific correction method against the benefits – for your particular situation. There are usually several valid solutions. We will give you the tools so you can select an optimum solution.

## What is the cost of balancing a rotor?

### If balancing is an added cost, what is the benefit?

We always want to avoid added cost and to justify added cost we need a corresponding benefit.

We make cutters sharp so that they will cut better. Is cutter grinding an added cost? How do you justify the cost of grinding or replacing cutters? We increase cutter speed to get higher throughput. We build machines more rigid to get better surface finish.

Should we run blunt cutters slowly on a lightweight machine to save the cost of cutter sharpening, high speed spindle drives and heavy iron castings? That would avoid all those extra costs.

What would be the quality level and production rate in this situation? We would have poor quality, low production rate and may not even be able to produce the parts of such equipment.

The question is not whether cutter sharpening is an added cost, but whether it is a required cost. Does the cost of cutter sharpening get recovered by the production of parts with less machining time and better tolerance?

It follows then that the question becomes "Is balancing a required cost?"

It is possible to manufacture rotating equipment without using balancing. What would be the impact on performance and reliability?

If bearing life is less than the warranty period, or customers are complaining about noise and vibration, where do we allocate those costs?

The point here is that NOT balancing a rotor can be very expensive.

Balancing is an operation that saves money by improving performance, reducing downtime, and improving customer (or operator) satisfaction. We need to understand the huge costs that we avoid by proper balancing of our products.

This is simple, right?

## What are alternatives to balancing?

We could ask ourselves "What is the cost of not balancing?"

The cost of not balancing is the expense of making the system sufficiently robust to meet all specifications, performance and reliability requirements; without resorting to rotational balancing.

When there is an unbalance, the result is that the equipment has vibration.

To fix vibration the machine can be made more rigid – that is an extra cost item.

The loads on the bearings from centrifugal force reduce bearing life, or we need to use bigger bearings – again this makes the machine more expensive.

To avoid the vibration and bearing life problems, we can drop the speed – but that equates to less production or lower performance.

Maybe we will have a surface finish problem – giving the perception of poor quality. In many applications the vibration also results in noise. The operator can also suffer due to the noise and vibration – you get tired more quickly and it is harder to concentrate in a noisy/high vibration environment.

Difference in surface finish due to balancing of grinding spindle.

Down time is expensive. A new spindle may cost US$2,500 but having the machine down for two days to fix it may be 100 times that. In the nuclear power or petro-chemical industries the failure of a US$1,000 pump can result in a system shutdown that could cause revenue loss in excess of a million dollars.

Refinery.

If we examine products such as inertial gyroscopes, high speed electric, motors, steam and gas turbines, reciprocating combustion engines, automobile wheels, etc., we find that many products we take for granted would not be practical without having balancing as part of the manufacturing process.

Clyde Barrow (Bonnie & Clyde), a well-known Ford driver, once wrote a letter to Henry Ford stating: "For sustained speed and freedom from trouble the Ford has got ever (sic) other car skinned."

The 1932 Ford V8 was renowned as the #1 getaway car and was also favoured by John Dillinger.

"For sustained speed and freedom from trouble the Ford has got ever (sic) other car skinned... Clyde Barrow"

So what was special about the 1932 Ford V8? It was the new, one piece cast block, V8 engine which was claimed to be 20 years ahead of its time. Incidentally it was also the first production vehicle to feature a balanced crankshaft which enabled smooth operation at a higher rpm!

As stated earlier, not all vibration is due to unbalance. We can only balance to correct for vibration at 1× operating speed. The image below shows vibration at several frequencies. The nominal speed of the pump is 1,800 rpm but under load it slows slightly. We have vibration at 1×, 2× and 4× rpm and at 7200 rpm (Electrical noise). This looks to be a good case for balancing which should fix the 1× vibration. It may also fix the harmonics if they were generated by the vibration energy.

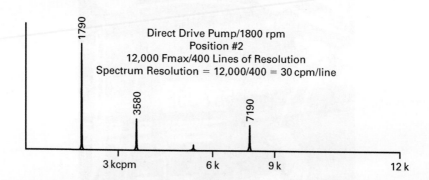

## What is the benefit of proper balancing?

One example would be a racing engine.

A racing car is a high vibration environment. Among racing cars nothing is more intense than Formula 1. This ultimate in high performance and high rotational speeds has pneumatic valves clattering up and down, pistons, connecting rods, combustion, gears spinning and vibration due to unbalance of the crank. Amongst all this one, can make a significant improvement in its overall performance by getting the crank balance right.

F1 engine – V10.

Back in 1998 Formula F1 racing engines were spinning only at about 14,000 rpm. I happened to be visiting a formula 1 engine manufacturer in England to look at their new balancing machine. A small machine that had cost about US$30,000, in a shop filled with multi-million dollar grinders and machining centers. While we were there the engine designer came by with a crankshaft and wanted to check the balance. When we looked at what he was trying to do it was obvious that this extremely competent and well educated engineer was ignorant on the subject of balancing (not uncommon!). We suggested a simple design change to the balance correction locations and to the balance tolerance. This was completed during our visit (less than 2 hours). After these simple changes the same engine was able to spin to 16,000 rpm – potentially the difference between first place and fourth place in the next F1 race. This crankshaft from an 800 hp engine weighed less than 15 lb (6 kg). In 2004 the engine technology had advanced to enable speeds up to about 19,000 rpm and I really wonder how much this could be increased with a little more attention to proper balancing.

Jaguar F1 racing car.

# Unbalance is not good for my product!

So, now we have looked at our product and come to the conclusion that spending some time and money on better balancing can actually save money; improve product quality; enable the performance data to be up rated; and have a higher level of customer satisfaction.

# How do we implement the program?

Since what we are looking at is a symptom, it is tempting to concentrate on alleviating the symptom. If short bearing life were the symptom then larger bearings would be a cure. If the symptom were excessive vibration then a more rigid mount would cure the symptom. This is not, however, the best approach.

If you visit the doctor and he says you have a high temperature (symptom) you will likely ask why. If the doctor replies 'Because you are sick' then get a new doctor. Do you need to find the reason why you are sick? Actually what you are interested in is just one thing – being cured! The cure needs to be based on the root cause of the problem or it won't be effective!

For machine vibration, we need to find the root cause – it is the unbalance. What causes the unbalance? Where do we start?

We need to remove the unbalance – that is so obvious. But there are only three options: to either add mass, remove mass, or redistribute mass.

We do this in one, two or sometimes more, axial locations.

A rotor is balanced by adjusting its mass distribution in relation to the center of rotation. There is a point when improving the mass distribution does not improve performance and so we stop balancing. In order to know when we should start balancing, and when we should stop, we need more information.

The level to stop balancing is when we have cured the problem. We could do some correction, run the system, do more correction, run the system again, and continue until the problem is gone. This is very labor intensive, not cost effective, and liable to introduce more problems related to multiple assembly and disassembly sequences. We could balance as low as we can go. Should we stop at 1 gram.inch? Should we stop at 0.1 gram.inch? What if we only needed to balance to 5 gram.inch? A lot of time is wasted balancing to a tighter tolerance than is needed.

What we have to do is find the worst tolerance that fixes the problem 100% of the time.

## Mission possible

Our mission is to understand what unbalance is and what it does. The best place to start is at the beginning – let's do some time travel and go back to the beginning.

Let's move back to around 300 years ago, to a man who liked to sit in the shade under his apple tree and think about the universe.

## Sir Isaac Newton

No one really knows if he really was hit on the head by an apple.

We do know, however, that he thought about gravity and falling objects, and that led him devise the 'basic laws of motion'.

Sir Isaac lived in a time of the three-class system. The *working class* who worked and yet had nothing. The *middle class* who

worked but who had a house with servants, and the *upper class* who were rich, had the house and servants but did not work. Isaac was upper class, had time on his hands, was smart, and did not watch television.

Isaac Newton did not have the benefits of computers and data analysis tools so he used the most powerful computer in the universe – his brain. We are all equipped with a 1.36 kg (3 lb) 'computer' (the weight of an average brain), capable of about 16 trillion calculations a second; unfortunately, most of us don't use anywhere near all that capacity.

Isaac Newton developed Calculus to solve some problems and showed what you can do with pencil, paper and a lot of time.

$$\int \frac{3x-1}{x^2+1}\, dx = ?$$

$$= \int \frac{3x\, dx}{x^2+1} - \int \frac{dx}{x^2+1}$$

$u = x^2 + 1$

$du = 2x\, dx$

$\frac{du}{2} = x\, dx$

$$= \frac{3}{2}\int \frac{du}{u} - \text{Arctan } x + C$$

$$= \frac{3}{2}\ln|u| - \text{Arctan } x + C$$

BACK SUBS

$32\, ft/sec^2$

$\Rightarrow$

$$= \boxed{\frac{3}{2}\ln(x^2+1) - \text{Arctan } x + C}$$

$\pi r^2$

PIE are round

Apple pie

Yummy

Pie in my tummy

Integration by Parts

$$\frac{d}{dx}(uv) = \frac{du}{dx}\cdot v + u\frac{dv}{dx} \quad \text{(Product Rule)}$$

Boldly go!

Now for the upper class, commercial trade was not considered proper; but working for the military was acceptable. Isaac Newton had a government contract to improve the accuracy of the British cannon. He looked at the latest high tech weaponry and its two problems – range and accuracy. In the course of this he came to the understanding that in one second the cannonball drops vertically about 5 meters (16 feet) and attains a downward velocity of 10 meters per second (32 feet per second). This is the same for a cannonball just dropping from the muzzle as for one leaving with a maximum velocity.

From this he theorized about a cannon on a very high mountain: a cannon ball fired fast enough (8,000 meter/s, or 7315 yards/s) that the earth dropped away from it as quickly as the cannonball itself dropped, resulted in a complete orbit of the earth. Sir Isaac came very close to the escape velocity, nowadays calculated with modern computers and proved with every satellite launched. From these principles he developed the basic theory behind planetary motion and centripetal acceleration.

Sir Isaac helped the army make major improvements in gunnery; hence the 'Sir' and he made a tidy profit.

From Sir Isaac's work (back in the seventeenth century) on gravity, acceleration and cannonballs traveling fast enough to orbit the earth, we now have current standard definitions and a clear understanding of the effects of rotation that we see manifested as vibration, force or noise.

## Newton's first law

Isaac Newton developed the three laws of motion which have never been improved on or changed. The first law of motion is that an object at rest will stay at rest, and a moving object will travel in a straight path unless acted on by a force. In the case of the cannonball, the force was gravity which caused a downward motion of the cannonball. This force is 'pushing' inwards and so is termed centripetal (inward).

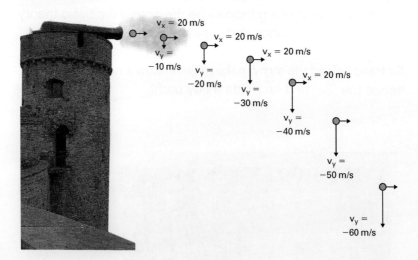

The fundamental behavior of all objects is covered in this law. In practice, there are many forces acting on objects. We can predict or measure most of these and determine the extent to which they impede our plans. To do this we must know the relationship between force and motion.

## Newton's second law

The second law of motion gives us the tool to calculate the forces. It is based on the simple relationship:

$$Force = Mass \times Acceleration$$

If an object is at rest and a force is applied to it, it will start to move, and the speed will increase. If no other force is applied then the direction will be constant, and the rate of change of speed (acceleration) will be constant.

This is the basis of space travel – apply a constant force for a period of time until the speed is high enough, then turn off the power, and keep going at the same speed.

In balancing we deal with a special case of this as we are dealing with a rotating object (the unbalance) that is traveling in a circular path. This means that for a given speed there is a constant inward acceleration. This in turn means that a centripetal (inward) force is acting on the unbalance.

We now need the third law to complete the relationship.

## Newton's third law

The third law states that every action causes an equal and opposite reaction. This is the key to understanding what we are doing by balancing a rotor.

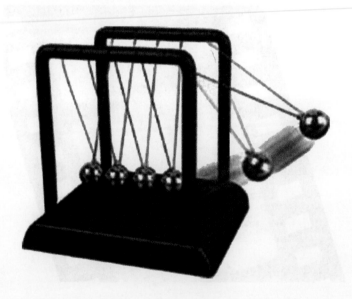

If an object is moving along a curved path the direction of its velocity is changing, and therefore a force is acting on it.

The force responsible for this change in direction is, as discussed above, called the centripetal force. It is directed toward the center of curvature.

By definition the unbalance is attached to a rotor. The centripetal force is reacted by an equal outward force which we are very familiar with. Centrifugal (outward) force is the reaction of the rotor/bearing system to the rotating unbalance.

If we are referencing the unbalance the force is inward – centripetal. If we are referencing the rotor then we are referencing the outward – centrifugal force. The unbalance causes a force on the bearings or a vibration of the rotor. When we balance a rotor we are compensating the unbalance mass to reduce the force and vibration.

## Energy balance

There is always a balance of energy input and energy output. It is easy to see the result of a single action, or a simple sequence of actions.

### Baseball

The Pitcher throws the ball (imparts energy to it). The Hitter swings the bat (putting energy into that). The Bat hits the ball.

The energy in the ball is reflected and most of the bat's energy is added to that of the ball resulting in it moving in a new direction with a new (higher) speed and energy level while the bat has a lower energy level.

## *Golf*

Another example is the golfer. He swings the driver giving it a huge amount of kinetic energy. The club hits the ball and most of the energy transfers to it, which then flies hundreds of yards (or not, depending on how hard it is hit).

Some energy may go into a divot and some remain in the club and absorbed in the follow-through action. To put the energy into the swing there had to be a reaction with the ground – that is provided via spikes in the shoes.

In both of the above examples there is a transfer of energy across an interface. In the case of a motor or pump the interface is the bearing and mounting system. Energy transfer indicates forces that have to be reacted.

The ball in both of these examples appears to be traveling with linear motion but Sir Isaac could easily calculate when gravity would bring it down. We now need to look at rotary motion, this is a little more complex.

## Rotary motion

With rotary motion we have the same law working.

The ball on the string is traveling at some speed. We know from Mr. Newton that, if left alone, the ball would travel in a straight line. As the ball travels in a circle there has to be an inward force applied to it.

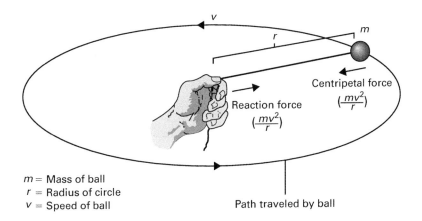

Centripetal force $(\frac{mv^2}{r})$

Reaction force $(\frac{mv^2}{r})$

$m$ = Mass of ball
$r$ = Radius of circle
$v$ = Speed of ball

Path traveled by ball

The force is called Centripetal – which we have already defined as an expensive foreign word for 'center seeking'.

Thanks to Isaac Newton we also know that to every action there is an equal and opposite reaction. In this case it is the tension in the string, which the hand feels as an outward pull – guess what we call that.

It is Centrifugal Force or 'center fleeing' if you don't want the expensive foreign word.

How would we balance this force?

Add another ball opposite, or cut the string?

In terms of balancing we replace the string with a disk. The ball becomes the unbalance which we correct by adding or removing mass.

Rotating disk with unbalance mass "*m*" at radius "*r*"

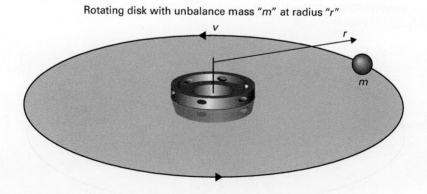

## Know your forces

As we learn about balancing we become conscious of situations where vibration and balance has frequently been ignored. You see a car moving along with one wheel bouncing up and down and you smile as you think 'unbalanced wheel'. Then you notice a co-worker's cooling fan shaking on the desk. You become attuned to these items and see what the cause of the problems must be. That does not mean you leap into action and fix it – you would have to work on every fan in the building!

Sir Isaac figured out the forces that were operating around him. Vibration is an effect but an unbalance force is the cause.

On the next page the image shows a balanced situation. Each of the 'airplanes' has a centripetal acceleration. Each one has a similar airplane opposite that applies an equal and opposite force. Each airplane has an average sized person. Essentially the system is balanced ('Balanced' is defined as being within acceptable limits) and the final result is close to zero.

---

**Problem**

If a couple of the 'airplanes' on one side of the carousel were empty and next to each other, while the others had people each weighing 175 lbs, we would have 350 lb of unbalance.

**Solution**

Move a person from the other side. In other words, redistribute the mass.

---

Guess what we do in balancing. We figure out where the heavy spot is and remove some material. We find the light spot and add some material. In other words, we move material from the heavy side to the light side.

## Inward or outward?

If the carousel airplanes (see above) experience centripetal acceleration, what do the riders experience? The answer is: Centrifugal force – they are pushed outwards. An observer on the ground sees only the centripetal acceleration. Centrifugal forces are only experienced in the local frame of reference (rider or foundation of ride).

Physicists call the centrifugal force 'fictitious', or virtual, but that is rather like saying that pain is not real – I cannot feel your pain, so it is not real to me, that does *not* mean that it doesn't exist.

## Flying free

Let's explore this example a little further.

What would happen if the chain that held one of the seats broke?

Where would the 'plane and rider go'?

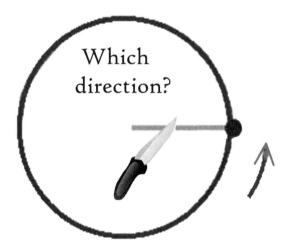

What we need is Isaac Newton's Law #1 "Every object will remain at rest or in uniform motion in a straight line unless compelled to change its state by the action of an external force."

## Which direction?

If you take away the force (centripetal) then the object will continue to travel in the same direction.

"Every object will remain at rest or in uniform motion in a straight line unless compelled to change its state by the action of an external force."

Intuitively we expect the object to fly outward since that is the direction of the force. The object has mass and velocity and therefore momentum. All the force goes into changing the direction. When the accelerating force is removed the object will travel in a straight line.

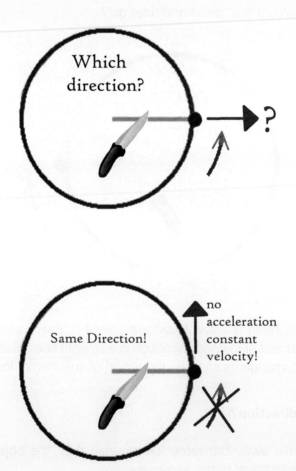

### *Bearing load*

The force to the bearings is the reaction to the tendency of the rotor to spin about its center of mass. The bearings constrain it to spin about the center of the bearings. If the rotational speed remains the same then the energy content of the object is constant.

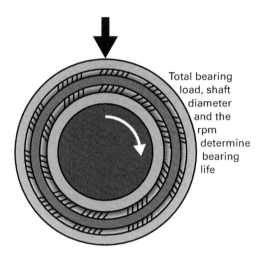

Total bearing load, shaft diameter and the rpm determine bearing life

The level of the energy content is based on the rotational speed and the radius. For a given rotation speed (rpm) the speed of the object is proportional to the radius.

For a given rpm an increased radius causes increased energy content. For the same energy content increasing the radius reduces the rpm (object has same linear speed).

Everything we do in balancing is related to the rotating mass, its radius, its speed and the resulting energy content. We are normally interested in the effective UNBALANCED mass only and not the entire mass of the rotor.

## Acceleration

Acceleration is the rate of change of velocity. Velocity is the rate of change of distance in a defined direction.

Late for school?

Rotary motion involves a continual change of direction and therefore a continuous acceleration even when the rpm is constant. It is this continuous acceleration of the unbalance mass that generates the vibration.

Acceleration is always a 'squared' function. Displacement is measured in meters or inches, velocity is measured in meters/second or inches/second and acceleration is measured in meters/second/second ($m/s^2$) or inches/second/second ($in/s^2$).

We can look at this in another way. Distance is a change of position with no time constraint (It is 8 kilometers (or 5 miles) from my house to the supermarket). However, to get from home to the supermarket I must travel in a specific direction. Whereas a distance of 5 miles from home could be anywhere on a circle with an 8 km (5 mile) radius; only one direction will lead to the supermarket. When we use the term 'displace-ment' it is implied that a specific direction is involved.

Citroen 2CV (1981) in low speed corner.

Velocity is a change of position in an amount of time, or the rate of change of position. With velocity there is an implicit direction. Speed is a directionless rate of change of position. If you are driving on the highway at 55 mph (88 km/h) the speedometer does not take account of direction. If you are traveling in a straight line, the speed and velocity will yield the same numeric number. If you change direction the speed may be the same but the velocity has changed. This is right back to the Newtonian laws.

 "Every object will remain at rest or in uniform motion in a straight line unless compelled to change its state by the action of an external force."

Just as velocity is the rate of change of position, so acceleration is the rate of change of velocity. Traveling down the highway at a constant speed and in a straight line there is no acceleration. When we come to a bend in the road and turn the steering wheel there is acceleration, the velocity is changing.

If as we travel we come to what Americans call a 'traffic circle' and the British a 'roundabout' it is possible to turn through 180 degrees and go back down the road in the opposite direction. The speed may be the same but the result is completely different. The result is actually the same as if we had braked to a stop, engaged reverse gear and accelerated back to the previous speed (except for the fact that we are not trying to steer looking back over a shoulder).

Just to ensure confusion, a European roundabout!

Let's look at what happened at the roundabout. After turning through 90 degrees our motion in the original direction had dropped all the way to zero but we now have motion at right angles to the original direction. At this point we begin to accelerate back to the opposite of the original direction.

What is the difference between a traffic circle and a roundabout? There is one significant detail. A British roundabout is traversed in a clockwise direction while an American traffic circle is traversed counterclockwise (for which the British term is 'anti-clockwise').

Our journey around the traffic circle illustrates what happens to an unbalance mass on a rotor.

When we are dealing with unbalance we have a fixed mass at a fixed radius of rotation. Once the rotor is spinning the unbalance is subject to acceleration, and therefore the centrifugal force is directly proportional to the square of the speed.

So where would we be without that guy called Newton? We need that second law of motion.

# Force = Mass × Acceleration

"Newton's second law states in part that the acceleration of a body is proportional to the resultant force exerted on the body and is inversely proportional to the mass of the body."

Translation, for a given force, the greater the mass the less the acceleration. A force of 1 Newton will accelerate a mass of 1 kg by 1 meter/s$^2$. If a force of 1 Newton is applied to a mass of 10 kg the acceleration will only be 0.1 meter/s$^2$.

$F = M \times A$. M is the unbalance mass. The velocity is (rpm.2PI/60) × Radius and the acceleration is Velocity × (rpm.2PI/60) – the term PI is converting the units from revolutions to RADIANS and there are 2PI radians in 360 degrees.

Let us now look at a special kind of force, the 'moment'.

## Moment of a force

A moment is not a short period of time: in this situation it is another word for 'torque'.

When a moment is applied to a mass it causes it to rotate about its axis rather than to move the mass center.

Typically, when we apply a force to a mass we are thinking in terms of moving the entire object. That is what happens when the force is applied through the center of mass. If the force is applied to the end, or an edge, of an object the center of mass may not move but the object may spin around its center. Some systems are designed for this. For example, a playground 'merry-go-round' or a revolving door. With other systems the effect may be unintended.

"Just how do we make this thing spin?"

## So which way is IN then?

"Am I coming out or going in?"

If we are tightening or loosening a nut with a wrench we are applying a torque. The rotational torque is called the 'moment' and it is the applied force (unit Pounds [force] or Newtons) multiplied by the radius arm (inches, feet or meters) and expressed in units such as foot-pounds or Newton-meters.

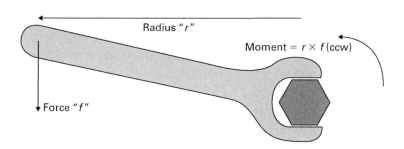

Radius "$r$"

Moment = $r \times f$ (ccw)

Force "$f$"

If we need to increase the torque there are two possibilities – either increase the force or increase the radius (use a longer wrench). **Note** to be correct it is the force perpendicular to the axis – pulling off axis reduces effective force.

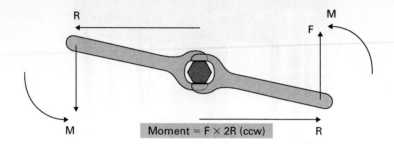

Moment = F × 2R (ccw)

OK, own up, some of you may have looked at the illustration above and thought the answer would be '2F × 2R'. Logically we have only doubled the moment, not quadrupled it.

An understanding of moments is important. Later on we will be considering moments when we deal with multi-plane balancing. When we have a condition known as 'Couple Unbalance' the result is a moment.

## Moment is a vector quantity

We have to consider not only the magnitude of the force but also the direction in which it is applied. Moments can be added or subtracted. Consider the above bolt with the two wrenches. Either we have two forces, each at length A, or one total force at length 2 × A. The rotational direction is the same so they add together. If you were to pull both wrenches towards you the forces would cancel out and you would have 'zero' moment or torque. This situation is equivalent to a 'Static Unbalance', it is effectively applied through the center of the object and tends to displace it rather than turn it.

## Examples

A wheel nut on a car must be tightened to 67 ft/lbs. The wrench is 20″ long to the center of the grip. How much force is needed (to nearest 1/2 lb)?

**Answer:**   Moment = 67 ft/lb = 67 × 12 in/lb
Moment arm − 20 inches
Force = 67 × 12/20 = 40 lb (to nearest $\frac{1}{2}$ lb)

This next one is not quite so easy. It takes a while to get the mental picture of what is happening.

A spring balance comprises an arm pivoted at one end, a suspension spring at 5 inch radius with a spring rate of 1″ per pound, and a pan at a radius of 8 inches. Adding a weight to the pan causes the spring to extend by 0.1 inches. What is the mass of the weight?

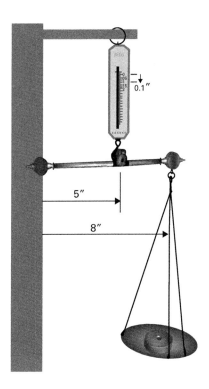

**Answer:**   Spring constant = 1 in/lb
For extension of 0.1" load = 0.1 lb
Moment = 0.1 × 5 = 0.5 in/lb
Moment at pan = 0.5 in/lb = 8 in/oz (OK little trick with numbers)
Radius at pan = 8 in
Force at pan = 8/8 = in/oz/in = 1 oz

## Relation of force to balancing

Balancing tolerances should be determined according to the loads on the bearings or the vibration of the system. In most situations we have a fairly rigid system, or sub system, and the performance and reliability is directly related to bearing loads.

If you increase the performance of a motor by raising the speed you may end up with vibration or bearing life problems. If the motor was originally designed for 1,800 rpm and the speed is increased to 2,400 rpm then the various components of the motor are likely to be stressed beyond their design limits.

Increasing the speed by 50% will result in bearing loads in a rigid mounting system increasing to more than double (225%) the original. To keep bearing load the same, the unbalance tolerance must be reduced by this amount.

# Free climbing rope
### (One per customer)

# As is, no warranty.

Nothing worth having comes free.

Since bearing life is rated by both speed and load then the load would have to be reduced by more than this amount to maintain the original bearing life estimate.

Nothing worth having comes free, right?

Why do we put so much focus on the bearings? The simple answer is that a high percentage of equipment failure involves the bearings.

In systems using electric motors the major maintenance work is replacement of bearings. This is partly due to the simplicity and reliability of an electric motor and partly due to the fragility of the bearings. Lots of things can cause a bearing to fail (too little lubricant, too much lubricant, contamination, impact loads, misalignment, high temperature, etc.) but when a motor is stripped and rebuilt the evidence for most of the problems is lost. The rotor unbalance is one of the items that can be checked. Rotor unbalance should be checked before re-assembly of a motor because operation and maintenance can cause the unbalance to change. If there is a large change then the cause of that should be investigated.

It's important to check the unbalance for the same reason that a climber checks his ropes carefully. You want to tackle a problem when it is easy to fix. The time to find your rope is frayed is not when you are halfway up a cliff!

## The cost of increased performance

Consider the situation of the motor with increased speed given above. We increased the motor performance by 50% but did we accomplish this for zero cost?

If the focus was just on shipping a functioning motor, one could argue that there is no additional cost. Maybe there was a different winding of the coils of the stator, or possibly a different phase angle for the commutator to coil displacement, but these do not necessarily increase the cost.

There may have been costs from using a different bearing or more expensive grease but these are minor changes.

The balancing operator just saw his tolerance cut and he needs two more measure and correction cycles on each unit to reach the new tolerance. Pity the poor production manager who just saw the balancing cycle time increase and cannot

## "Our policy on bad news? Shoot the messenger!"

produce the required number of motors per shift. Everything in this motor is now more highly stressed, and the production team is also highly stressed! It is not uncommon to have design changes for higher performance (or lower manufacturing cost) that result in higher failure rates, warranty returns and customer complaints months after the change.

What usually happens is that production operations get the blame. If one of the problems is vibration then usually the balancing tolerance gets cut (again).

The easy answer is always to tighten the balance tolerance. If that does not work then tighten it more. Unfortunately that still may not cure the problem.

The dilemma is that we have a system that was designed for a set of performance criteria but the requirements were changed. This is such a common occurrence that it is seldom considered.

# 18 Mb Hard Disk Drive!

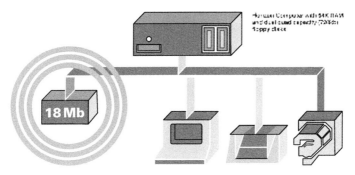

High tech disk drive circa. 1986.

In the computer disk drive industry this cycle of design change was incredibly rapid for a few years. In the far off reaches of history (around 1980, which is prehistoric in the computer industry) computer hard disk data storage involved huge and massive systems with limited storage. The disk themselves were 12″ (35 cm) or more in diameter, held a few megabytes of data, needed a clean room operating environments and specially built balancing machines to achieve the exacting tolerances. Once development of 'consumer' disk drives started then production volumes soared and new companies and drive models cropped up almost monthly.

Standard disk drive for home/office PC.

One of the design objectives was to eliminate balancing (to reduce both cost and particle generation). Once balancing was removed successfully, data density was increased and balancing was designed back in. Then the cycle would repeat. Each new generation needed more balancing machines since the volumes were increasing. Now the industry is mature, drive with hundreds of gigabytes are inexpensive and design evolution is by 'use of best practices' to refine the design in small steps to be more cost effective, reliable and with fast data rates. In the peak of development (1990s) high tech, robotic machines for disk drive balancing were the high point of balancing machine development for products that had a life cycle of 2–5 years. Today disk drives are a commodity available at discount electronics stores. Balancing tolerances have been refined to be as wide as possible without compromising data integrity or product life.

## Balance tolerances and performance

Once tightened a balance tolerance is not likely to get relaxed. If the tolerance is relaxed and 6 months later warranty failures start to increase then the person that loosened the tolerance is in the spotlight. Of course, if nothing happens there is no glory since no one notices anything. It is not like painting the motor a different color – everyone sees that. So where is the incentive?

There is a benefit to getting the tolerance right. A looser tolerance means shorter cycle times, less material removal or addition, and therefore lower production cost (it may also make the difference between needing one balance station or two, so avoidance of capital expenditure and need for a second operator may figure in here).

It takes guts (and often some capital investment) to be the one to relax the tolerance. Data is needed to prove that the result of the change will not be higher costs later. For example, an end-of-line vibration test with frequency analysis. This can reveal the truth – is there significant 1× rpm vibration. In this situation one must also review factors such as assembly variations, bearings, and fans that may have been added after balancing.

Equipment like this CAN reveal what is happening but only when operated by someone with the knowledge and skill to interpret the readings. Naturally, one can hire a consultant, with experience in balancing and vibration, to bring in equipment, run tests and give qualified advice – always an excellent idea!

Vibration can be generated in many ways. Unbalance is one of the main causes of vibration but bearings, fans, gears, mounting, assembly techniques and many other problems can also generate vibration. Knowing what to look for and how to evaluate the data is crucial to fixing the problem.

We can review the performance aspect in another way. By the addition of a suitable end-of-line (EOL) test system you can detect not only the occasional high unbalance rotor, but also the more common bad bearing or incorrectly mounted bearing. The EOL test can also find items such as missing spacers or loose components which can be fixed before shipment to reduce warranty problems and improve customer satisfaction.

## Unbalance and speed

Balancing tolerances are often derived from the ISO 1940 spec-ification. This is based on a measurement of vibration velocity – this is what we are sensitive to when we touch something that vibrates – it is a combination of frequency and amplitude. Units are inches/second or mm/s.

A large amplitude at low speed feels the same as a smaller amplitude at high speed. There is always an exception – when the dentist comes at you with that drill the vibration velocity doesn't matter!

The choice of units here is important.

Measuring vibration velocity in units of mm/sec gives the numbers for the ISO 1940 quality grades (G6.3 is 6.3 mm/s). The graph shows speed at the bottom and the corresponding vibration amplitude (0 − peak) to give 6.3 mm/s.

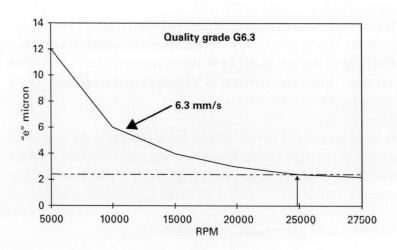

If we wanted to use inches per second the equivalent number would be 0.248 in/s.

This relationship is based on the specific condition that the rotor is spinning in free space without restraint and the unbalance is located in the plane of the mass center. The rotor will then rotate about its mass axis, the bearing axis will be parallel to the mass axis and the distance between the axes is the eccentricity while the velocity of the bearing axis is the quality grade number in mm/s. In theory the vibration amplitude is the same as the mass eccentricity under the above condition. In practice, this is never the case. The ISO grade gives a starting point to determine the optimum value for the specific application.

The ISO 1940 standard assumes that the bearing supports have 'zero' stiffness. Now consider the opposite case.

## Centrifugal force

The calculation of centrifugal force is actually quite easy if you have rigid bearings and no vibration – a theoretical situation, like a physics problem with massless beams and frictionless rollers.

There is a precise mathematical relationship between unbalance speed and centrifugal force.

Don't worry, this isn't a math class so we are not going to do a bunch of problems. What we need to know is the relationship.

To calculate the force the formula is:

$$Fc \text{ (pounds)} = U \text{ (oz.in)} \times (rpm)2 \times 1.78 \times 10^{-6}$$
$$Fc \text{ (Newtons)} = U \text{ (g.mm)} \times (rpm)2 \times 1.1 \times 10^{-8}$$

Two things are important to remember.

1. The force is proportional to the unbalance.
2. The force is proportional to the SQUARE of the speed.

## Unbalance and speed

Many people think that unbalance gets worse as speed increases. We know that this is not the case.

This graph shows that unbalance is a constant – it is a mass property of the rotor.

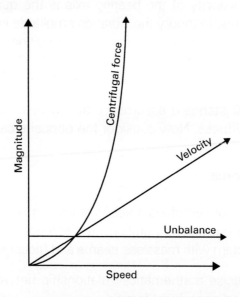

If a rotor with 'zero' unbalance is run at any speed there will be no vibration or bearing forces. This is a theoretical situation only, of course.

If a weight with a mass of 1 gram is added to the rotor at a radius of 100 mm then when it spins there will be vibration and bearing forces that will change with speed but the unbalance is fixed at 100 g.mm – unless we spin fast enough for the weight to approach the speed of light when it would get shorter and more massive – but that is another problem.

We do consider that the rotor is rigid. More than 90% of all balancing situations have rotors that are rigid. There are certain rotors that are designed to flex in operation but these are a special case.

A long slender rotor, such as a drive shaft, will bend at high speed. This is not a change of unbalance but a rotor deformation problem. The unbalance tolerance and unbalance correction locations have to be selected to ensure safe operation for the operating speed range.

As speed increases vibration velocity increases in direct proportion. Velocity is displacement multiplied by frequency. The eccentricity (displacement) is a constant, while the frequency increases.

Centrifugal force follows a square law relationship since it is the rate of change of velocity = velocity times frequency.

As speeds increase, the bearing load increases by speed squared. Bearing life is based on load times speed, so bearing life is inversely proportional to speed cubed. A speed increase of 25% with the same unbalance cuts bearing life in half (assumes a completely rigid mount system).

Here is the root of the balance tolerance difficulty. We can calculate vibration against tolerance with zero stiffness bearings and we can calculate bearing load against tolerance for infinitely stiff bearings. The real situation is somewhere in between, and for a given manufactured assembly the bearing stiffness can vary over a wide range depending on how the unit is installed.

These unknown conditions lead us to work in a situation where the actual bearing situation is ignored. If we use mass eccentricity we have a useful unit, analogous to specific gravity, that is independent of the mass of the rotor.

### Unbalance defined

By definition a rotor has an axis of rotation. For (say) a crank-shaft the axis of rotation is defined by the main bearings. For a flywheel it is defined by the combination of the bore of the mounting hole and the shaft on which it mounts.

## V6 CRANKSHAFT

A rotor has a mass axis which passes through its center of mass – often called center of gravity. Due to manufacturing errors and material variations the mass center does not line up with the geometric center – tolerances are the definition of how much error is allowed!

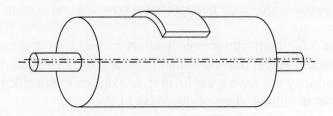

When the rotor spins the distance of the mass axis from the rotational axis is the eccentricity of the mass – this is a radius. Multiply the radius by the rotor mass and the result is the unbalance.

$$U = m \times r = M \times e$$

## Everything together

The illustration below shows all the parameters for unbalance.

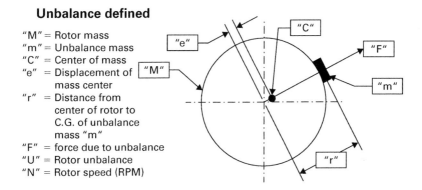

**Unbalance defined**

"M" = Rotor mass
"m" = Unbalance mass
"C" = Center of mass
"e"  = Displacement of
       mass center
"r"  = Distance from
       center of rotor to
       C.G. of unbalance
       mass "m"
"F"  = force due to unbalance
"U"  = Rotor unbalance
"N"  = Rotor speed (RPM)

Always use Mass instead of Weight because we are dealing with centrifugal force not gravity. Even a satellite floating in orbit has to be balanced so it does not wobble on its axis and the unbalance has the exact same effect with or without gravity.

For a balanced rotor the mass is located at the bearing center (on the bearing axis). When an unbalance is added, the mass of the rotor is moved away from the bearing center. Unbalance defined as mass × radius is the same as specific unbalance defined as the rotor mass × eccentricity.

If you are driving along the road and a bug hits the windshield the speed of the bug changes by a large amount while the speed of the car changes by a small amount. The energy transferred to the bug is the same amount that is transferred from the car, but the bug has lower mass. A small unbalance mass at a large radius is like the bug.

---

### Example

A balanced rotor weighing 20 kg and with a diameter of 2 meters has an unbalance of 1 gram added to the outside diameter. How much does this move the mass center?

$$m \times r = U = M \times e$$
$$1 \text{ gram} \times 1,000 \text{ mm} = U = 1,000 \text{ g.mm} = 20,000 \text{ g} \times e$$
$$e = 0.05 \text{ mm}$$

---

## Correcting unbalance. How do we do that?

Unbalance is corrected by adding, removing, or moving mass. These are the only three options.

Repeat until the rotor is within the required tolerance. **Note**: not to 'ZERO' [Nothing has a zero tolerance].

People who would ask you to 'get all the unbalance out of the rotor' would also call you crazy if you were to ask for a tighter tolerance on shaft surface finish, diameter and circularity. By definition this is a 'measure, correct and audit' process. The output depends on the input. On one day 90% of parts may not need correction, and the next day 90% of the parts may be way off.

---

**NOTE**

The process of balancing involves redistributing the rotor mass but the end parameter we are working on is actually ensuring that the noise, vibration or bearing life are within design parameters.

# 3

# The balancing process

## Overview of real balancing tolerances

What are the limits for actual balancing operations?

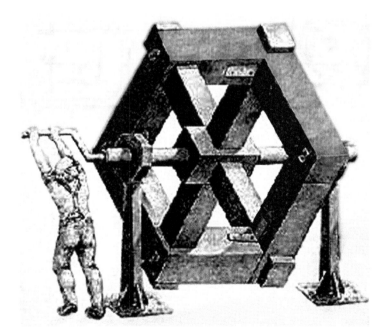

Precise Manufacturing and balancing is important.

Ten microinches (0.25 micron) of mass eccentricity is about the practical limit of measurement with a standard balancer. For special applications we can reach 4 to 5 microinches (0.1 micron). Be careful – machines can measure to lower indicated levels than that, but the repeatability issue will get you, as if you run the rotor again you will not necessarily get the same reading.

In balancing we use numbers that are much smaller than those used in most machining operations. A balance tolerance may well be equivalent to ten microinches when the surface finish on the journals is 60 microinches or more. If the machine is capable of measuring to 4 microinches then the difference between two consecutive readings can be 4 microinches. A micrometer may be able to measure to 0.0001" (0.0025 mm) but repeated measurements with multiple operators (and possibly multiple calibrated micrometers) will give a range of values that can vary by up to 0.001" (0.025 mm). Resolution is not the same as accuracy or repeatability.

With any measurement system dealing in units of 100 microinches (2.5 micron) or lower there are inherent measuring problems. As an example consider the 'simple' measurement of a shaft journal diameter. First, it is temperature dependent since the shaft (like all metals) exhibits thermal growth. Second, there is a question of whether the journal is truly circular (it could be oval and so has a large diameter 90 degrees away from a small diameter or it could be tri lobed, which shows that the journal has a constant diameter, but the mating bearing will not fit over the shaft). Third, is the variation between measuring tools (calibration). Finally, we have measurement repeatability which is both operator and equipment dependent.

Let us use as an example a steel shaft with a 3" (75 mm) diameter bearing journal which has a tolerance of 0.0003" (0.008 mm). The machinist could get a measurement in the shop at the lower edge of tolerance (winter time). But three hours later the QA department, in a warmer lab and with a different measuring instrument and measuring location, could reject the part for being oversize.

For the measurement of unbalance there are additional factors. We cannot physically access the mass center of a rotor to measure its location but we have to spin the rotor on its journals and derive the unbalance value from the synchronous loads on the support bearings. Everyone involved must have a high level of confidence in the stability, repeatability and accuracy of the balancing machine.

Until about 1995, the standard test rotor design for aircraft turbine engine balancing was a long tube with heavy ends, and fitted with double row ball bearings at each end. The design was developed around 1950 to maximize the moment of inertia and match it to the characteristics of the soft bearing balancing machines with 'three degree of freedom' work supports (vibration front to back and swivel in two orthogonal directions). By 1985, the industry had changed to hard bearing technology with much greater machine sensitivity and accuracy. It was soon apparent that this test rotor design had major problems. The first problem was maintenance of the double row ball bearings. It could take days to remove, clean, reassemble and retest just to get the bearing performance stable enough to pass the test. The second problem was the instability of the long tube section of the rotor.

> I was personally doing a test in the early 1990s when a loading bay door was opened about 100 feet away from the balancing machine and a draft of cold air blew across the shop. When the cold air hit the rotor (spinning at 900 rpm) the balance readings went off the scale and did not settle down for several minutes. The new design of rotor was short, solid and with plain journals. This works very well but even so the tester must be careful not to rest a hand on the rotor when moving test masses since that small thermal input can change the readings enough to cause test failure.

The situation is worse for flywheels, fans and other rotors that mount to another shaft in service. For rotors without their own shaft there is an issue of tooling. Very seldom can we repeat the location of tooling better than about 200 microinches – 0.0002" (5 micron). It doesn't matter what you calculate the tolerance to be; in practice the limit of balancing is the repeatability of the tooling when the rotor is removed and replaced randomly.

There are two additional problems once the rotor is balanced. The fit between the rotor and the service shaft and the runout of that shaft can cause unbalance in service conditions that is several times higher than the actual unbalance of the rotor.

---

**NOTE**

We are not talking about the sensitivity and accuracy of the balancer, which can easily be 10 times better than the best tooling, but the real accuracy of the balance level in a rotor such as a flywheel. In real terms this is the true vector difference between one measurement and a subsequent measurement.

For rotors with shafts that are to be fitted with rolling element bearings there is another issue. There will be runout of the inner race of the bearing that has a direct effect on the balance of the assembly.

## Summary

Given the need to balance a rotor for satisfactory operation under actual service conditions it is not enough to set a balance tolerance from a chart or specification. The balance tolerance must allow for what happens after balancing and that can mean that the tolerance needs to be lower than would be calculated by such specifications as ISO 1940. Another consideration is that subsequent conditions may cause such large changes that it makes no difference to balance to a tolerance two or even three times higher than the specifications would indicate (in these conditions it may be required to trim balance the assembly).

In determining the balance process the need for tooling is an important consideration and the design of that tooling to achieve the required repeatability is a critical factor in the success of the project. Just as there are many applications where the optimum balance tolerance has been determined by testing and evaluation there are many where the tolerance was taken from another product or from a chart and is probably inappropriate for the real needs.

## Rotor unbalance defined

A rotor with a single unbalance. This illustration also appears on page 96 where we discussed a similar concept.

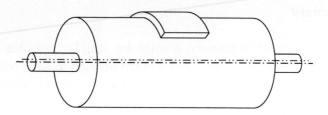

How easy it would be if rotors came like this and you could just knock off the unbalance.

Normally you cannot see the actual unbalance; it is the total of a number of small manufacturing errors.

One of the problems with balancing is that you cannot directly see or measure the unbalance, only the effect of the unbalance. You cannot see electricity either but the effects can be just as dramatic if you ignore either one. Did you ever receive an electric shock? Did you see it coming?

To represent the sum total of all the unbalance of the rotor we show a single mass at a specified radius. $U = m \times r$. This defines the magnitude of the unbalance.

---

**NOTE**

The unbalance is located at a specific angular location which must also be known or there is no way to make correction.

---

If we just need to know if unbalance level is 'OK' or 'NOT OK', we can ignore the angle in this case.

## Specific unbalance

The unbalance mass and radius causes a corresponding shift in the position of the total mass of the rotor.

Unbalance $U = M \times e$

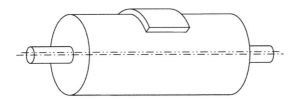

In many cases the unbalance arises from mounting errors such as bore clearance or shaft runout. If the rotor is (say) a flywheel mounted on a motor shaft then there has to be a clearance between the shaft diameter and the flywheel bore. This clearance can cause the flywheel to be mounted off center. This mounting error is an eccentricity (e) and the mass is the rotor mass (M).

Mounting error has the exact same effect as unbalance. The flywheel and the motor may both have been balanced accurately but the mounting error may cause unacceptable vibration.

Individual rotor assemblies may be rejected at final testing due to system vibration. When they are put back on the balancer they test as in tolerance. So why was there excess vibration?

This is a process problem and arises from the lack of understanding of the effect of small amounts of runout, bore clearances and assembly variations. In the realm of balancing an
error of 0.001" (0.025 mm) can be a major problem.

## Sources of unbalance

When a rotor assembly reaches the balancing machine it has
normally been the subject of a number of machining and
assembly operations. Each of these operations can contribute
to the total unbalance of the rotor.

### Where do balancing errors come from?

**Example, an armature from an electric motor.**

Shaft straightness
Laminations – bore location,
thickness variations
Pressing laminations on shaft
Commutator
Installing commutator
Windings
Impregnation
Turning the OD.
Bearings

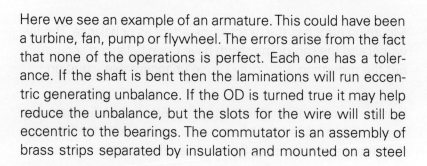

Here we see an example of an armature. This could have been
a turbine, fan, pump or flywheel. The errors arise from the fact
that none of the operations is perfect. Each one has a tolerance. If the shaft is bent then the laminations will run eccentric generating unbalance. If the OD is turned true it may help
reduce the unbalance, but the slots for the wire will still be
eccentric to the bearings. The commutator is an assembly of
brass strips separated by insulation and mounted on a steel

hub. Again there are several ways for the commutator to be out of balance.

The eccentricities, clearances and distortions add together in random ways so that a batch of supposedly identical rotors can have a wide range of initial unbalance values.

The balancing process is not a single operation but a 'measure, correct, audit' cycle.

## Force, unbalance and speed

As you may have realized by now balancing is FUN^2.

$$F \propto UN^2$$

We now understand that the main reason for balancing is to reduce the bearing load and bearing vibration to acceptable levels. Balancing also produces a reduction in noise due to the reduction in vibration. Unbalance robs power from the system that is dissipated in the form of vibration and noise and can cause other problems like stator rub and early bearing failure.

The unbalance does not change with speed but the effect does. Centrifugal (outward) force is proportional to the square of the speed.

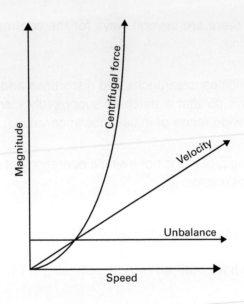

The balancing machine measures the amount and angle of the unbalance so it can be corrected by adding, removing or moving mass.

We use mass since we are not talking gravity (apples falling on heads) but rotational conditions that are the same on Earth, Mars, Jupiter, or in deep space.

Balancing is the correction of the rotor mass distribution of the rotor relative to the defined rotational axis.

# How to balance?

## What is balancing?

A rotor that is out of shape won't run smoothly.

It is obvious that there is no way that a rotor with misaligned bearings like this would run. It is unbalanced so we need to bring it into balance by aligning the rotor mass with the bearing centers. The original version would not run. It would destroy bearings and shake loose everything around it.

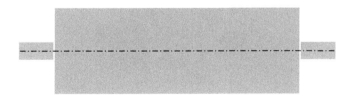

Guess what, that is exactly what happens when the errors are a few ten thousandths of an inch – a couple of hundredths of a millimeter. All this fuss about balancing is about errors that rarely go above the 0.005" range down to less than a ten thousandth of an inch.

If you attempted to build a high speed machine with tolerances close enough that you could eliminate balancing, it would cost a huge amount and may even not be possible to make. Balancing is not an added cost item but a process that enables other operations to be less precise and therefore the product is less expensive to produce.

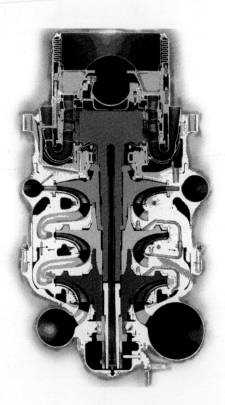

Balancing is an enabling technology that has been the key to navigational gyroscopes, jet engines, turbo pumps for rocket engines (like the space shuttle) and even the automobile engine. The high speed machining of airplane parts is dependent on high precision balancing and the machinery making 'Facial Quality Tissue' could not run without very sophisticated balancing operations. Just consider for a moment that tissue paper is produced at a speed of around 7,000 feet per minute in a ribbon 20 feet wide. It is very fragile, especially when wet. The consequences of not balancing are mind boggling!

## Rotor balancing

The illustration below shows a typical aircraft engine rotor mounted in a horizontal balancing machine. The shaft is located on rollers. There is an end thrust roller to restrain the axial motion – note that is mounted to the upper part of the machine. There is a safety shroud with interlock switch. The shroud keeps the operator from touching the spinning rotor and retains objects such as balancing weights (the object is called a weight and we classify it by its mass). The shroud also reduces the amount of drive power needed and avoids generating airflow that would mess up the operator's hairstyle. Not visible in the picture is the drive belt around the shaft to spin the rotor.

This setup enables us to spin the rotor so that the unbalance will generate centrifugal force reactions at the bearings and result in signals produced by the pickups (transducers). In other words we get signals we can filter to show unbalance amount and angle. Once we know that we can correct the unbalance.

**Rotor axes**

Here we have a rotor.

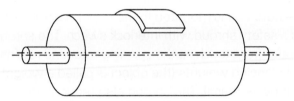

It would mount in service, and in a balancing machine, by the bearing journals. In this case we do not have to worry about tooling, etc. The rotor has a bearing axis defined by the center-lines of the two bearing journals. It has a mass axis defined by the rotor mass distribution. Here we see a symmetrical rotor but with a large additional mass on the OD at the axial center-line. This will displace the mass axis parallel to the bearing axis (same amount and angle at each end).

If we sat the rotor on a pair of knife edges what would happen? That weight would drop to the bottom.

Gravity will pull the heavy side down.

## Review definitions

The rotor is symmetrical except for the unbalance m at radius r

$$U = m \times r = M \times e$$
$$U = M \times e \quad \text{so} \quad e = U/M = m \times r/M$$

The unbalance mass × its radius equals U. Divide this by rotor mass and we get 'e', which is a measure of unbalance that is independent of rotor mass. It is called mass eccentricity, or specific unbalance. It is the displacement of the mass center from the bearing center.

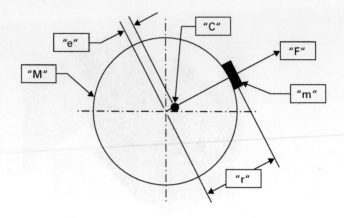

## Unbalance defined

"M" = Rotor mass
"m" = Unbalance mass
"C" = Center of mass
"e"  = Displacement of
        mass center
"r"  = Distance from
        center of rotor to
        C.G. of unbalance
        mass "m"
"F"  = force due to unbalance
"U" = Rotor unbalance
"N" = Rotor speed (RPM)

An unbalance of 1 oz.in (720 g.mm) may be insignificant for a power generator rotor but would be catastrophic on an electric drill. A mass eccentricity of 5 micron is the same for both rotors.

## Mass axis

If we took the rotor up in the space shuttle and launched it spinning like a satellite (spin stabilized) it would rotate about the mass axis.

The earth is in perfect balance. It spins about its mass axis. When we move goods, services, people, etc. about the planet we change the mass distribution and the earth changes its spin axis to stay in balance (this does not imply that individual people are in balance).

When we force an object to spin about a defined bearing axis there is a fixed reference for the rotation axis (bearing axis). If the mass is not evenly distributed about that fixed axis then we have unbalance. The axis about which the mass is evenly distributed is defined as the mass axis.

The mass eccentricity 'e' is the measure of the unbalance in terms of the displacement of the mass axis and the bearing axis. Units are linear – inches or mm.

Here we are not using 'eccentric' to imply people acting strangely. The eccentricity multiplied by the rotor mass gives the unbalance. The units are the combination of mass and eccentricity – ounce.inches or gram.millimeter (note the multiplication function).

In this situation we are looking at the measurement of unbalance – since we have not defined a correction location we cannot reference that yet.

## What is a micron?

Unbalance and eccentricity are interchangeable. We need to get used to the numbers we use here.

Let's put these numbers in context. The outer circle represents the typical diameter of a human hair 0.0035", about the same as the thickness of a sheet of paper. Next we have a circle diameter 0.001". Pretty well the normal limit of repeatability for any

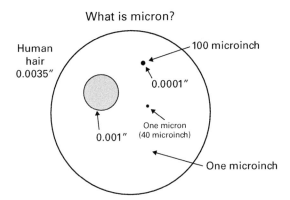

What is micron?

mechanical type of tooling. With new hydrogrip tooling you can get somewhere between 0.001–0.0001. We have a dot for 0.0001"–2.5 micron (The Hydrogrip system uses a high pressure grease gun to expand or contract a thin steel membrane).

We can usually balance down to a micron without much trouble – until you move the rotor on the tooling! Of course, if you are balancing something that has its own shaft that is not a problem.

Finally, there is 1 microinch – this is so small that you can't see it! We can't even measure it unless we use a high speed machine capable of spinning the rotor at 10,000 to 100,000 rpm.

## Balancing defined

The rotor wants to spin about its mass axis but we want it to spin about the bearing axis. The results are force on the bearings, vibration of the bearings or a combination of both.

---

**NOTE**
More rigid bearings mean less vibration and higher force.

---

Our definition is taken directly from ISO 1925, paragraph 3.1. The ISO 1925 definition of balancing is:

> A procedure by which the mass distribution of a rotor is checked and, if necessary, adjusted in order to ensure that the vibration of the journals and/or forces on the bearings at a frequency corresponding to service speed are within specified limits.

Note the progression – check against limit, adjust if required. Note also the important phrase "corresponding to service speed". Balancing cannot fix vibration at other frequencies than the rotational.

## Force/vibration

The point that is made in ISO 1925 is precisely "that for a given unbalance condition the vibration is inversely related, and force on the bearings is directly related, to the bearing stiffness". Unbalance is not shown by the external appearance but by the vibration or force it generates. A rotor may have a hole drilled in the outside. That may be there to correct the unbalance rather than being the cause of the unbalance.

How do we know the wind is blowing – we feel the breeze.

How do we know if a power socket is live?

The electrician asked his new helper to pick up a cable and touch the end. He said "do you feel anything?" The helper said "no." The electrician said "whatever you do don't touch the other cable – it has 440 volts on it!"

It is better to use a meter or light bulb to check if there is high voltage! The balancing machine is the meter for unbalance. Rather than putting a hand on a motor, feeling the vibration and deciding if it feels OK we need to have a way to make real measurements of the rotor condition.

## Rotor rigidity characteristics

There are a number of classifications of rotors, depending on flexibility, operating speed, and other factors (ISO 5343).

**Class 1** is rigid rotors – this covers 90% of all applications.

**Class 2** is rotors that are not rigid or that have special characteristics of mass distribution but that can be balanced using a modified balancing technique (choice of correction planes is the key here).

**Classes 3 and 4** are flexible rotors.

We mention this so that you are made aware that some rotors may have to be balanced at specific speeds, at two speeds, or even when hot. Thermal effects can cause distortion that in turn causes unbalance, which can cause more distortion.

Some call this type of operation 'dynamic straightening' rather than balancing since the aim is to keep the rotor running true by minimizing internal bending stresses of the rotor.

Not a project for a novice.

## Static unbalance

Static unbalance is defined as acting through the mass center of the rotor and by definition can be corrected at a single location by adding or removing material in line with the center of mass. Implicit in this definition is the fact that the mass axis remains parallel to the bearing axis. Static unbalance can be detected without spinning the rotor, hence the name.

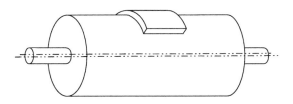

Balancing on knife edges can correct for static unbalance but without any qualitative measurement of the remaining amount of unbalance – depends on friction, wind, shaft dia, rotor mass, etc. This does not confront the basic definition of checking the unbalance against a known standard. Most single plane balancing is done on rotating (centrifugal) balancing machines. Non-rotating (gravitational) balancers are used for high volume production of rotors with coarse tolerance. These measure the calibrated deflection of a spring or the offset force using two load cells.

## Couple unbalance

Couple unbalance is what you get when you balance on knife edges and don't correct the real source of unbalance. The image below shows a rotor that had an unbalance at one end and was static balanced with correction of the end opposite the unbalance. Result is that it looks good on knife edges but will shake like crazy as soon as it spins. Note that the mass axis crosses the geometric axis at the center so there is no static unbalance.

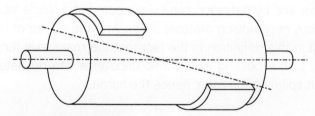

Rotor with coupling unbalance.

If you have ever wondered about the reason for looking at moments, now you know why.

'Couple' is a moment, a twisting force. The magnitude of the couple is dependant not only on the mass and the radius, but also on the distance between the masses. If you moved them so that they were opposite (zero distance) they would cancel each other out since they have the same mass and radius. The measure of couple is the mass times the radius times the distance from the CG. Units therefore are ounce inches squared or $g.mm^2$. On a narrow rotor the bearing forces are small as the bearings are much further apart than the unbalance masses. On a dumbbell shaped rotor couple unbalance is a killer, as the bearings see a multiplication of the unbalance because they are close together, while the unbalance masses are far apart.

For overhung spindles couple is not a major problem as the couple bending moments are absorbed in the shaft and the bearings are widely spaced. However, this is beyond the scope of this book.

For couple unbalance we need to consider the locations where unbalance arises in conjunction with the bearing locations. In many applications couple unbalance can be high compared to static unbalance as the forces are internal to the rotor and do not affect the bearings.

## Dynamic unbalance

Dynamic unbalance is the generic combination of static and couple unbalance. Real world rotors start with unknown unbalance – it is cute to make nice pictures but rotors don't come like that. The illustration on the next page shows the corrections needed to bring the rotor into balance.

The key thing with dynamic unbalance on a rigid rotor is that all of the random unbalances throughout the rotor – runout, holes deeper than others, voids in material, variations in Gibbs etc. can be combined and resolved into two correction masses at predetermined axially separated locations.

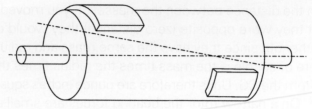

Rotor with dynamic unbalance.

Vector addition of the unbalances gives a result that represents the forces (or vibration) at the bearings and can be mathematically translated to the appropriate masses at the correction planes. Let's repeat that in English – within reason we don't care where the unbalances actually are, we pick correction planes and relate the unbalance to those planes. The tolerances are related to the forces at the bearings but the correction is related to the correction planes.

Measurement of the unbalance first tells us whether it is within tolerance or needs correction. If correction is needed we need to know how much material has to be added or removed and we need to know the axial and radial location to add or remove material.

## Measurement and correction

Let's consider a typical rotor.

The next image could be of an armature, a pump rotor, or a printing roller. The configuration is with two bearings and two correction planes (A and B) inboard of the bearings. An unbalance anywhere on the rotor will generate forces at the bearings. The bearing forces can be classified by amount and angle using direct measurement. These measurements can be translated into corresponding amounts and angles at the correction planes.

It is easier to see this in reverse by looking at an added unbalance and relating this to the effect on the bearings.

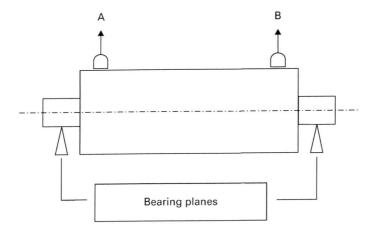

## Single unbalance – 1

Next we have a balanced rotor with an added unbalance of 10 units at zero degrees. With the dimensions shown, 80% of the unbalance will load the left bearing and 20% on the right bearing. The angle will be zero at both ends. From this we see that if we measured the unbalance as 8 on the left and 2 on the right at zero degrees we could calculate the correction needed as 10 at 180 degrees located 20% of the bearing distance from the left bearing.

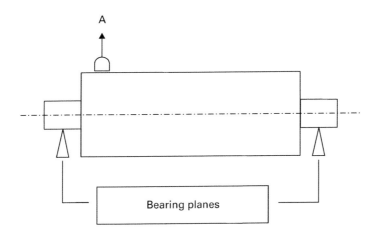

## Single unbalance – 2

Now look at a different unbalance condition. We have unbalance B of 6 units, in the center of the rotor at an angle of 90 degrees. The arrow direction illustrates this. It should not be hard to see that this will give an unbalance of 3 units at 90 degrees at each bearing. This is Static unbalance. Unbalance in plane of center of mass.

What was the previous condition? Was that also static unbalance?

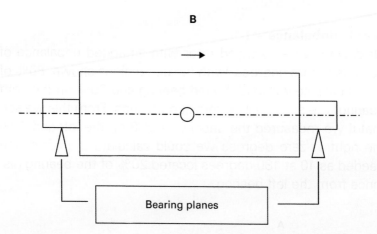

Bearing planes

## Single unbalance – 1a

Looking at the next image, we see a single plane unbalance and could be corrected with a single correction mass – right. What would happen if we tried to correct the unbalance in the center at 180 degrees? We would have a couple unbalance. What should we do to correct this unbalance with a single mass? Move the correction plane to the left!

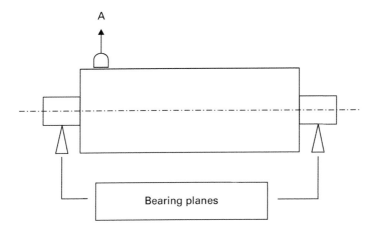

This is not static unbalance as it is not symmetrical, but it can be corrected in a single plane – this is called QUASI STATIC unbalance. The lesson from this is that if a rotor assembly goes through an operation that changes the unbalance at one axial location, then we can trim balance using a single correction plane. Consider a motor that is fitted with a drive pulley. If the rotor is then out of vibration specs we know it was the pulley that caused it and can therefore make a single plane unbalance correction on the pulley.

## Single unbalance – 2a
Back to our static unbalance. Look at the first illustration on the next page – is this static unbalance? We could correct this with a single mass in the center or a half sized mass at each end. This situation usually arises with a disk shaped rotor such as a pulley, axial fan or flywheel. It is simpler to explain when we use the same rotor for all illustrations.

What would we do if we had a rotor with both of these unbalances?

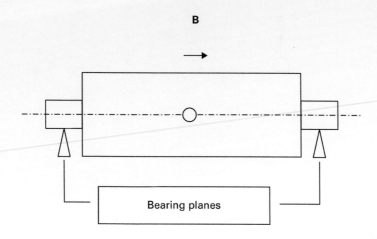

## Two unbalances

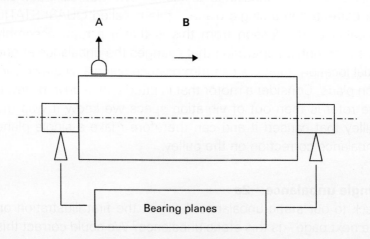

We need to add the unbalances at each end.

So what do we have on the left 8 + 3 = 11 = does that look right? NO! We have 8 at 0 degrees and 3 at 90 degrees on the left, and on the right 2 at 0 degrees and 3 at 90 degrees.

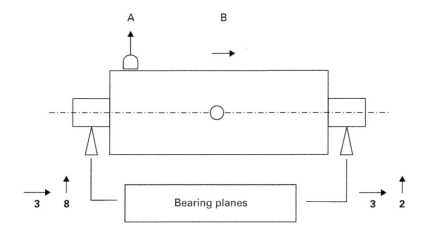

We have to add the vectors. Which would you rather do, get your calculator and do math or draw a picture? I vote for the picture.

## Vector addition

Let us look at the result of two unbalances on a rotor.

## End view of rotor and unbalance effects at the bearings

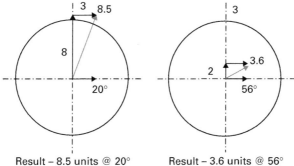

Result – 8.5 units @ 20°          Result – 3.6 units @ 56°

From this we see that an unbalance of 8.5 units at 20 degrees can be resolved into its components of 8 units vertically and 3 units horizontally. We also see the importance of correcting unbalance at the right angular location. An unbalance of 8.5 corrected by 8 units with a 20 degree error does not leave 0.5 units but 3 units. We will come back to this later.

For now we need to understand that we can translate the effect of an unbalance at one location in terms of a different unbalance at a different location but with the same effect. What we just did was take a balanced rotor, add two unbalances and find out the effect on the bearings. On a balancing machine we do the reverse operation – we measure the unbalance at the bearings and translate that to the effective unbalance at the correction planes.

## Unbalance in two planes

"With two correction planes you can balance any rigid rotor"

This general case of a rotor with unbalance and two correction planes covers 90% of all balancing applications.

What varies? Bearing locations, correction locations, single plane or two plane. With two correction planes you can balance any rigid rotor. With a single correction plane you can balance narrow rotors or rotors where the axial location of the unbalance is known. For example if we balanced this rotor then added a pulley to the end any unbalance would arise from the pulley.

## Unbalance outside the bearings

Below we have a rotor and pulley and the pulley has an unbalance of 10 units at 0 degrees.

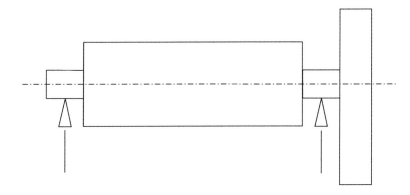

Since the pulley is outside the bearings the effect of the unbalance is magnified on the right support and negative on the left – the total is still 10 units – 11 + (−1) = 10.

We have two obvious methods of correction.

We can do a two plane balance with correction on the rotor or we can do single plane correction on the pulley.

The advantage of correcting the pulley is that the rotor stays in balance! Assuming the rotor was balanced before addition of the pulley the assembly will be in balance if the pulley only is corrected.

---

**Balancing tip**

Choose the simple solution whenever you can. Life is already too complicated.

---

### Getting to be too much?

What is needed is a method to reduce one plane without affecting the other. This can be done mathematically with the use of calculus or six simultaneous equations. Another way is to use a balancing machine with the calculation system built in.

We don't want to have to make calculations every time we use a balancer. What we need is for the balancing machine to do the calculations for us. Here is a typical measuring screen showing the unbalance corrections that need to be made.

This screen does not show the rotor geometry (we only need to see that during setup).

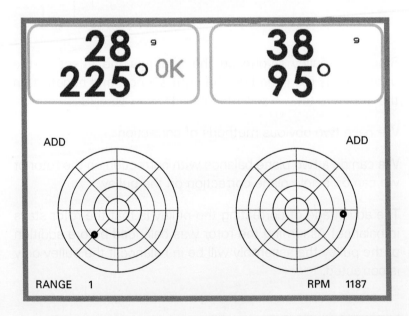

### *Plane separation*

For a balancing machine to be easy to use we need a simple way to translate the data. The data comprises signals from the pickups and the phase reference. From this we know the rpm, the 'zero' angle and the relative amounts and angles of the unbalance at each bearing. But we don't usually correct the unbalance at the bearings! The other half of what we know includes the dimensions of the rotor and locations where unbalance correction is possible.

Here is the measure screen with the setup information overlaid.

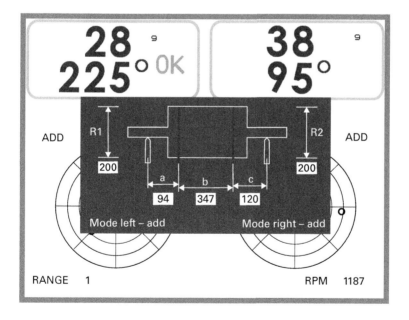

**Review what we know**

Location of 'zero' angle, locations of bearings, radius where we can correct, axial locations where we can correct. Basically the rotor dimensions that we can measure with a ruler.

### Machine setup

At the top of the screen is the unbalance data for the setup conditions below. On modern equipment, changing the setup data will automatically update the screen display without having to run the machine again. A closer view of the setup shows the data.

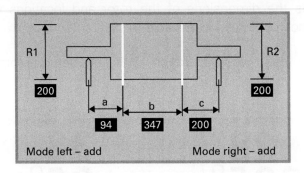

We see here the rotor picture – in this case a standard "between the bearings layout and three dimensions highlighted". At each end of the rotor picture the diameter for correction has been entered (the machine converts this to the appropriate radius number). We have the measurement 'a', this is the distance measured from the contact points of the left support rollers to the left correction plane. Next we have the correction plane spacing 'b'. Finally the distance from the right correction plane to the contact point of the right support rollers is 'c'.

Once the five data values are input, the balancing machine computer will be able to convert the signals received from the pickups directly to read the actual correction required at the two correction planes. Since we have the radius the machine can make the conversion from gram.mm to grams, ounces or another convenient unit.

The balancing machine is concerned with the ratios between the 'a', 'b' and 'c' dimensions so they can be in any convenient unit – provided they are all in the same unit.

From a practical point it is best to use the largest numbers that the machine allows. For example, with 'a' = 1, 'b' = 2, 'c' = 1 the only options to make an adjustment of the 'a' dimension are 0 or 2. If the respective numbers are 100, 20, 100 then 'a' can be adjusted to 99 or 101 which is 1% of the value. With analog systems this was more critical since the potentiometers used to set the values had a certain amount on non-linearity which was amplified when using only 10% of the maximum value (if the unit had a full scale error of 1% then the actual error in the value could be 20% if the slider was at the 5% position).

### Balancing machine basics

The same system is used on analog machines. The illustration (on the next page) shows a typical small hard bearing analog instrumentation balancing machine.

Unbalance readout is by means of two amount meters and two angle meters.

For machine setup there are five potentiometers a, b, c, R1 and R2. These set the correction plane and radius dimensions just as on the screen of the microprocessor instrumentation.

Dial in the numbers you measure with a ruler and the machine is set.

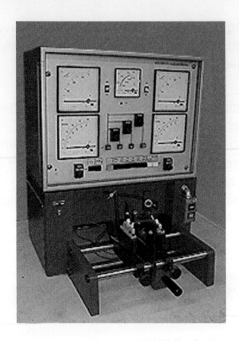

## Measure, then correct unbalance

Once the machine is set it will provide direct readout of the unbalance for the selected planes. Either the rotor is in tolerance, needs correction or is uncorrectable (opposite).

The amount reading indicates how much unbalance there is at the selected correction plane. It is important to remember that we need to correct to bring the rotor into tolerance, rather than to 'Zero'. Removing more material costs time and increases the chance of damaging the rotor performance. For example, if we are balancing electric motor armatures, then material removal can affect the motor torque. Microprocessor systems with a polar graph give a direct visual readout of the action required. As you can see, there is unbalance but we are in the 'bulls eye' so no action is needed. The rotor is in tolerance. If the tolerance is set correctly balancing lower will not improve performance.

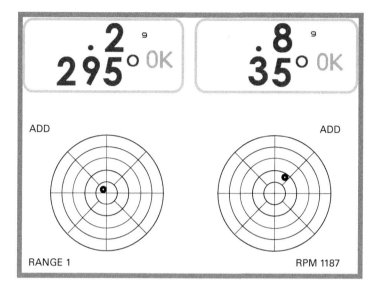

## Correction

First we need to set up the balancer. Once the balancer is set the rotor is loaded and measured. When it is not within tolerance we have to make some correction. Up till now the process has been simple, the machine is designed that way.

Correction is often difficult to achieve. Designers do not always make (adequate) provision for unbalance correction when designing rotors. Sometimes there is no way to remove or add enough material. Sometimes the available material is not in areas that are effective (small radius, close together, not separated from mass center). Sometimes the problem is with the precision that is needed in mass removal – an example would be freehand grinding where operator skill is critical for efficient operation without ruining parts.

Unbalance correction is where there is an immense amount of variability. Corrections may be just a few milligrams on a small high speed rotor or could be many kilograms or a large ship propeller. Balancing often takes place on almost finished assemblies and the appearance after balancing is important. Production throughput may mean there is only a few seconds to make the correction. In many cases a production balancing system can involve a multistation transfer line.

In order to have an efficient correction of unbalance we need to break the problem down into a series of smaller decisions.

## Correction, 3 ways

(Sounds like a menu item in a Chinese restaurant).

There are only three choices:

***Choice #1*** – Add mass – bolt on a block, add a screw, add a washer under a screw, mix some epoxy putty and apply it or weld a block in place. Watch for: obstruction that will prevent rotor from turning. Make sure weight can't fly off and don't weaken the rotor drilling and tapping a hole to apply a weight.

***Choice #2*** – Remove mass – drill, mill, grind, remove balance weight (maybe a washer under a bolt). We need to watch for: cutting too deep, weakening part of rotor that may distort at operating speed, damaging windings on armature, and leaving chips on the bearing journals that could damage machine or rotor at next measuring run.

***Choice #3*** – Redistribute mass – asymmetric rings that turn, vary length of bolts holding pulley on shaft, rotate the pulley on the shaft or open up an undersize bore with a radial offset to correct unbalance (mass centering). Watch for: overtightening of adjustable pieces, having too much offset in eccentrics (limits precision), or having too little offset (limits maximum correction). It is critical to ensure that adjustable pieces do not move in service since this can cause large unbalance or result is contact with non-rotating components with catastrophic results.

In looking at a specific rotor we first decide which of the three choices is most appropriate. Within that choice we further break down the options and look at production volumes.

Some other considerations include housekeeping and safety. If material removal is being used we may have chips that need to be vacuumed away. With any cutting tool there is the need for safety shrouding to protect the operator.

## What are the limits?

We need to establish a couple of reference points.

There are International standards that we can use as guidelines for balance tolerances but they are by nature very general. Relying just on one of these standards can give you a tolerance that is quite a way off from the optimum.

---

### Example

API 610 paragraph 2.8.4.10 –

"Pumps shall operate smoothly while accelerating to their rated speed. Variable speed pumps shall operate smoothly throughout their speed range"

Even the ISO standard suffers from a bad case of the 'fuzzies'. The introduction reads as follows: "The recommended balance quality grades are not intended to serve as acceptance specifications for any rotor group, but rather to give indications of how to avoid gross deficiencies as well as exaggerated or unobtainable requirements."

## Noise is the limit

When a machine is running there is a background vibration at many different frequencies. Some of this comes from bearings, maybe a drive belt – especially one that is worn. Sometimes motors have a magnetic unbalance.

Fans may have a blade that is out of line. Gears have runout. All this leads to a total vibration signature.

Signatures, like fingerprints, are unique. Like fingerprints, signatures can be decoded but instead of handwriting analysis we use frequency analysis. Vibration measuring and analyzing equipment can break down the vibration into a spectrum showing the amounts at many different frequencies. This spectrum is what we call a 'signature'. Different problems cause effects at different frequencies. For unbalance the frequency of interest the running speed.

When we are balancing we are looking at only one frequency – 1 × rpm. When the synchronous vibration is lower than the surrounding vibrations, there is no gain from balancing lower.

### *Noise signature*
We now look at two vibration signatures.

The first is a representation of the vibration level filtered to look at one frequency at a time and then all the results plotted.

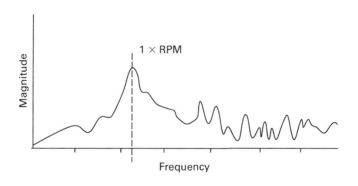

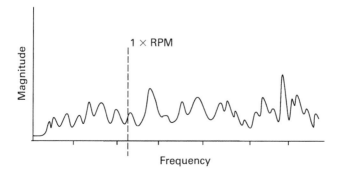

This first chart shows high vibration at 1 × rpm. This means it is likely that there is significant unbalance. There can be other causes for this vibration but we are looking specifically just at balance right now.

The second chart shows significant vibrations. Since the vibration at running speed is lower than the vibration at other frequencies there will not be much improvement by further balancing.

There is still unbalance and we could reduce the unbalance further. It is not the capability of the balancing machine that is the limit. The thing here is that balancing lower will not improve the noise, vibration or quality of the product.

## The right tolerance

We are looking for the tolerance that is most cost and performance effective. It will be close enough to enable full performance without premature bearing failure or excessive vibration. It will be loose enough to enable cost effective manufacturing.

Unfortunately, whenever there is a vibration problem the knee jerk reaction is to cut the balance tolerance in half. If that does not work then cut it in half again. Since the risks of a loose tolerance are warranty failures there is always reluctance to loosen an achievable tight tolerance. Only when it becomes unachievable does it get looked at.

The proper way to determine the ideal balance tolerance is to take a batch of well-manufactured rotors, balance them as

low as possible and then run them and progressively add unbalance until the performance is just acceptable. Now the unbalance can be measured and a limit determined that gives good results without being too tight. These rotors should also be checked for vibration signature under operating conditions to classify what other vibrations may affect performance.

This may seem like a lot of trouble but the result is worth it. First we have the lowest cost balancing tolerance, second we have the basis for the tolerance documented and finally when there is a change in rotor design there is data to evaluate the effect on balancing. Also there is no need for a guillotine!

One other variable needs to be considered in the production process – the variations that occur after balancing due to assembly and stack up tolerances. Armatures are balanced and then bearings are added. The motor is fitted with a drive pulley or a flywheel that has a loose fitting tolerance. The result is high vibration on a percentage of the assemblies.

Don't fall into the trap of blaming unbalance when the problem is a machining error or purchase of low quality parts. That guillotine may come in handy after all.

## Standards again

In order to help with determining tolerances, standards such as ISO 1940 were developed. This ISO is based on the measurement of machinery vibration velocity. The ANSI spec is identical but printed by American National Standards Institute. The API specification is written around pump requirements in the Petro-Chemical Industries and classifies unbalance levels as a function of rotor mass and operating speed.

ARP documents are built around the requirements of aircraft engines. They specify balancing machine requirements so that machines can be approved as meeting the needs of the industry while each manufacturer can set their own requirements.

Each of these standards organizations is partly funded by sales of these standards. The technical committees that draft them have to meet somewhere, the notes have to be formatted and mailed out and these costs show up in the price of purchasing the standard. They are copyrighted and should not be copied. It is permissible to use short extracts as we have done here.

## ISO 1940

ISO 1940 is famous for its classification of vibration in terms of G codes although many people don't know what they mean it is easy to figure out that G2.5 is a tighter tolerance than G6.3. Notice the choice of words here, tighter not necessarily better. G2.5 means a vibration velocity of 2.5 mm/s under specified conditions. Unfortunately, it is the theoretical value assuming the rotor was spinning in free space so it does not relate to actual operating conditions.

ISO 1940 uses a set of criteria to classify the acceptable vibration grade – a low speed marine diesel has a coarse grade while a high speed grinding spindle has a very tight grade. The tightest grades require balancing a rotor in its own bearings and under service conditions.

### *Example grades*

This excerpt from the examples given in ISO 1940 shows the typical types of rotor that fit into grades G6.3 and G2.5.

> The coarse end of the range covers items such as slow marine diesel engines, which have a grade of G4000. The fine end of the range is G0.4 which is for precision grinding spindles and gyroscopes balanced in their own bearings after final assembly.

Automobile crankshafts fit into Grade – G40 and Agricultural machinery into grade – G16.

In general, balancing tolerances are getting tighter. This is partly because it is possible and partly because equipment is getting lighter – or more performance is demanded from a given size of product. Some applications follow a cycle where balancing is designed out to reduce cost. Performance is increased but balancing is needed to obtain the improved performance and so balancing is designed in.

## Getting at the numbers

Here is the shorthand way to calculate a tolerance for your rotor.

$$\text{Using ISO 1940,}$$

$$\text{Balance Tolerance} = \text{(gm–mm)}$$

$$\frac{9.54 \times \textbf{G number} \times \textbf{mass} \text{ (grams)}}{\textbf{RPM}}$$

If you want the grade to be G6.3 with a 50 kg rotor at 1800 rpm the tolerance would be about 1,700 g.mm. If the rotor has a diameter of 300 mm (rad 150 mm) the tolerance would be 11 grams at the OD.

For two-plane balancing the tolerance would be 5.5 grams per plane.

If you wanted to change the operating speed to 3,600 the balance tolerance would drop to 2.7 grams per plane for the same vibration level. To maintain the same bearing life would require a further reduction or bearing re-design.

## API 610

API 610 is based on a very simple formula:

$T = 4W/n$ with rotor weight (W) in lb, Speed (n) in RPM and the resulting tolerance (T) in oz.in

A simple formula does not guarantee easy balancing. For high speed pumps, especially the tolerance can get ridiculously tight (when the tolerance is below 20% of the diameter tolerance of the bore it can be classed as ridiculous).

The drive trains in the petrochemical industry are often complex and can generate vibration which manifest in another piece of equipment by means of coupled pipes, common foundations, etc.

One very specific problem with the API specification is that it does not make any allowance for balance tooling error and bore clearance on the arbor. The idea is to balance to almost 'zero' on the balancer in order to allow for assembly errors. The result is that a high vibration pump may be stripped down, and the impeller balance checked only to find high unbalance (actually due to non-repeatability of mounting). After rebalancing and re-assembly there may still be high unbalance or, conversely, it may be OK. A secondary result may be that the operator loses confidence in the balancing machine and tooling.

If you work in a refinery you may be stuck with meeting the API requirements. You should at least make sure the tooling is good and fits well (also compensate the tooling unbalance and runout – see later) so that your hard work is reflected in low vibration of final systems.

It is always good to check the balancing machine setup data, such as minimum measuring speed, measuring mode, tolerance, etc. and adjust as needed; a previous operator may have used inappropriate settings.

## Fundamentals of balancing

### The importance of ANGLE

This section will explore the importance of the angle in balancing. If there is one problem in balancing that appears minor, but has major effects, it is angle error. It is usually easy to see when an unbalance correction is too small – the unbalance amount reduces and the angle stays the same. If the correction is too large then there is a 180 degree phase change.

When there is an angle error the amount may be over corrected, or under corrected but there is a major angle change that generates confusion. Often the operator will complain that he is "chasing the unbalance". After unbalance correction the amount is slightly reduced but the angle is very different. We have to remember that unbalance is a vector quantity and that means we have to know what?

# Magnitude and Direction!

## Which of these is more important?

This is a Trick question – they are equally important – let us see why

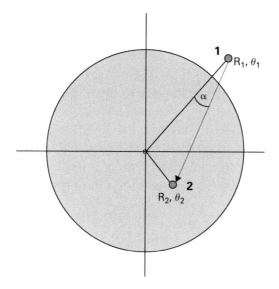

## *Phase error*

In the illustration above we have a situation where the unbalance was apparently at an angle of about 45 degrees. After correction the amount was smaller but was now at 140 degrees. What happened?

This may not be immediately obvious but the actual amount of correction was very close to perfect. The proper amount of material was removed. However, it was removed at the wrong angle. To be precise there was a slight over correction of about 5% but that was not the problem. The problem was that the unbalance was actually at 55 degrees and so there was a 10 degree error in angle of correction. Ten degrees out of 360 is less than 4%, right?

Wrong. We have to look at the dreaded trigonometry and evaluate the sine and cosine. If you have a scientific calculator you just punch a couple of keys and get the numbers for any angle. For the angle error of 10 degrees shown in the illustration above the cosine of the angle gives us the error in correction amount due to the angle error (the correction was less effective due to the error).

The Cosine of 10 degrees is 0.98 so no significant errors (2%) in the amount of correction due to the angle error (The cosine of '0' is 1).

The sine of the angle error gives us the value of the error at 90 degrees to the correction. This is a new unbalance generated by making the correction at the wrong angle.

The sine of 10 degrees is 0.17 so the amount error at 90 degrees to original unbalance is 17% of original unbalance.

### *Phase shift*
Looking at this more carefully we see the original unbalance (U) and correction (1) that yields the final result (2).

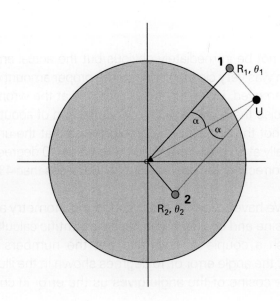

The correction was within the 5–10% error we normally expect with amount and due to normal error is things like drill depth, weighing error, etc. What this does is raise the angle change to more than 90 degrees. If we had made an under correction the angle change would have been less than 90 degrees.

The angle error is about 10 degrees and that means that the amount would have been under corrected by the Cosine of 10 degrees, which is 0.98 – or a 2% error. At 90 degrees to the unbalance the result will be the sign of 10 degrees, which is 0.173 or 17%.

With an accurate correction we would expect the unbalance to drop by a factor of 10 due to normal measuring and correction errors. If the angle is not calibrated correctly then the errors are much greater.

## Correction efficiency

With a system that is working well and has the capability of calibrated correction (e.g. drilling or adding weights) we expect to reduce the unbalance by a factor of about 10:1 in one shot.

The initial unbalance U is assumed to be a point but if we make a second measurement the value would change

**Correction ratio**

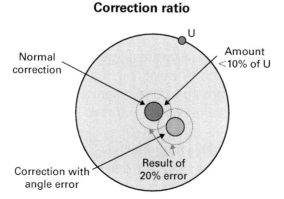

slightly. After correction the unbalance should be within the dark circle.

If we have an angle error (photocell moved, for example) we will have that 10% residual shifted by the SINE of the error angle. This would put the residual unbalance within the white circle.

If correction is made by a non-calibrated method (hand grinding, for example) the correction error will be larger and the effects of the angle error will be harder to identify. If we have measuring repeatability problems the effect will be the same. The dotted lines show the result of a 5:1 correction ratio. Note that the circles overlap, within that region it is indeterminate if the error is due to a wrongly calibrated angle or to correction errors.

Larger percentage errors are likely when working to very tight tolerances, balancing at very low speed or when there are other problems (bad surface on journal). This can make it hard to determine the real problem – actually the first step is to improve repeatability. This may mean calibrating the angle with a larger unbalance to give smaller percentage errors.

The single most common error on a belt drive balancing machine using a photocell to pick up a rotor mark is a wrongly positioned photocell.

To position the photocell correctly follow the manufacturer's recommendations. With variations in types and sizes of units there are several possible correct setups. Once the photocell is responding to the reference mark we can check the angle.

Run the machine with a rotor that has low unbalance and then add a large unbalance exactly in line with the 'zero' mark and half way between the correction planes. It is possible to use two test masses (one at each end of the rotor) but unless it is a test rotor there could be an angle difference in the test mass position.

The machine should read the same angle for each correction plane (there is only one unbalance). The photocell should be adjusted and another measurement made until the machine reads to remove mass at zero degrees. For safety reasons it is not recommended to adjust the photocell position with the machine running. It may be necessary to remove the test mass and rebalance the rotor and repeat the exercise to get an exact result. If the angle reading in two-plane mode is not the same for both planes then the machine needs an angle recalibration. When the angle reading is correct it is much simpler to make effective amount corrections.

## Unbalance vector

With a precise unbalance amount correction the residual unbalance is due to the correction being made at the wrong angle.

The amount of the error is directly related to the angle error – or more precisely – to the SINE of the angle error.

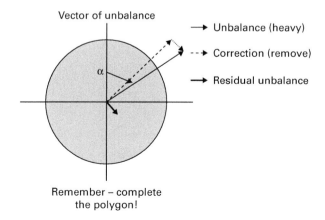

If you look at the illustration on the previous page you will see that the angle movement is in the opposite direction to the error. If the correction causes the angle to increase then the machine readout for angle should be increased. For example, the figure shows a correction made at about 20 degrees when the unbalance is actually at 30 degrees and the angle moves to 140 degrees. An angle error of 10 degrees does not result in an angle shift of 10 degrees!

**TIP** An angle shift of less than 90 degrees means that the unbalance was under corrected. When the angle shifts by more than 90 degrees it means that the unbalance amount was over corrected. Amaze your friends, look at the results and state confidently whether too much or too little material was added or removed – just look at the angle shift.

## Angle error chart

Here is a chart to correlate angle error to amount error.

| Angle error | |
|---|---|
| Angle | Amount |
| 1 | 0.02 |
| 2 | 0.035 |
| 3 | 0.05 |
| 4 | 0.07 |
| 5 | 0.09 |
| 6 | 0.1 |
| 7 | 0.12 |
| 8 | 0.14 |
| 9 | 0.155 |
| 10 | 0.175 |

There is a direct relationship between angle error and the effect of angle error reflected in the residual unbalance amount. At 10 degrees the amount error is 2%, which is still small compared to an error of more than 17% due to the 10 degree angle error. Angle errors of less than 3 degrees are acceptable for most applications. If the angle is critical, due to high correction ratio or limited available material for removal, then an electronic protractor (remote angle readout) will improve performance. This system tracks the rotation of the rotor after the measuring run so that is can be positioned with an accuracy of 1 degree or so.

**TIP** An angle shift of 90 degrees shows that the correct amount of material was removed but at the wrong angle. If you can remember that 5% is 3 degrees and 10% is 6 degrees (15% is 9 degrees) you can instantly predict the correction angle error. If the initial unbalance was at 30 degrees and the residual unbalance is at 120 degrees then add the error angle (in the same rotational direction) and make a further correction (at say 126 degrees). Even better would be to move the correction by 6 degrees (unfortunately you can't move a drilled hole).

### Correction ratio

We have mentioned correction ratio several times. Now we need to study this more closely.

In order to balance a rotor we first have to measure the unbalance. Once we know the unbalance magnitude the question is – is it in tolerance?

# In tolerance – no action

# Out of tolerance – make correction

After unbalance correction we have to make a check run to verify that the correction was done properly. An important measurement of the effectiveness of the measure and correction cycle is – correction ratio. Correction ration is very important when the initial unbalance is high (more than five times tolerance).

This is simply the ratio between the initial unbalance and the final unbalance after a single correction. With a well-adjusted measuring station and correction system it is normal to be able to achieve a correction ratio of 10:1 (when initial unbalance is high – 5× tolerance or more).

If the initial unbalance is high, and there is enough material available to add or remove, the correction ratio can be much higher than this. With semi-automatic and fully automatic machines we can sometimes see ratios between 20:1 and 50:1. If there is any problem with the rotor that affects measuring repeatability, or correction accuracy, this can reduce the correction ratio below 5:1.

When initial unbalance is low the percentage variations in the readings will be higher and the correction ratio will be lower. This does not matter if the rotors still come within tolerance after one correction.

Since production processes vary over time the average and peak values of the initial unbalance can vary from day to day and therefore the correction ratio is not the only parameter to use to classify the balancing operation.

In order to quantify the machine performance what we need to do is verify the measuring station by doing a Urr (Unbalance Reduction Ratio) test. Since this uses a test rotor and test weights it is a repeatable standard. Once the measure station is known to be accurate and repeatable the correction station can be set up.

To calibrate a correction station we have to use it to make changes to the rotor unbalance and correlate the changes in unbalance to the correction levels. We normally start with a rotor that is balanced to the lowest achievable level and make progressively deeper cuts at the same angle to generate setup data without making too many rotors unserviceable.

## Why not 'zero'

There are three variables that restrict the machine perform-ance; measurement error, correction mass, and angle error and plane separation. The last of these is not just a measuring station setup parameter but also a significant correction parameter (does the station act at the specified axial position).

One of the factors that separate balancing machines from most of the other processes is that each rotor is unique. Most production operations do the same thing each time. In balanc-ing each correction is a different amount and different angle compared to the previous rotor. With two-plane balancing there is also the relative amount and angle between the planes to consider. Rotor number one may have the unbal-ance angles at 30 degrees and 50 degrees while the next may have the angles at 30 degrees and 210 degrees and the cor-rection ratios can be very different.

## Measuring errors

The first reason for a poor correction ratio is that the machine does not measure with adequate precision. This may not be the fault of the measuring system of the machine! The setup and operational parameters may be to blame.

In a high production environment the measure time may be too short for accurate data collection. Maybe the drive belt is too tight, or damaged. An end drive machine may have a drive shaft that is too heavy or in poor condition.

Another reason for problems may be that the machine is not calibrated properly and has amount or angle error.

On smaller rotors the mass of the reference mark may be significant and re-applying a mark (dirty, damaged, or moved) may be a problem (sometimes a piece of tape is applied for measuring and then removed – it may be fitted at a different location next time).

## Correction error

The first reaction is to blame the measuring machine. However, it may not be the measuring machine that is at fault – often it is the correction that is the problem. The correction device (say a drill press) may be set up to drill at the wrong radius or the angle calibration for the correction may be wrong. Perhaps a new drill was fitted and the zero depth is wrong.

Sometimes an armature may have varnish lumps on the outside of the lamination stack that reduces the depth of milling (varnish weighs less than steel). A trapped drilling chip or build-up of dirt can have the same effect. An operator may not change a tooling nest when changing rotor size, and the rotor is not located properly in the nest – so suddenly there is a lack of repeatability. Tooling wears with use and we become accustomed to gradually reducing performance. Check tooling wear regularly. A blunt cutter may not take out enough material – or the machine may have been adjusted to compensate and then a new cutter installed so the machine overcompensates.

These are all common errors in production shops. Repair and maintenance situations also have error problems – although different problems.

## Plane separation

Plane separation problems can be due to the measuring station, the setup configuration or to a correction problem. The first action is to ensure that the measuring station is correct (with test rotor). A test rotor is designed specifically to enable a simple check of the measure station operation, accuracy and repeatability. Without accurate measurement nothing else will be accurate. The second action is to test the setup – are the correction planes too close or at poor location where perhaps there is insufficient material? The third action is to verify that the correction machine is set for the correct planes. The fourth action is to verify the average correction location compared with the maximum and minimum corrections. Sometimes unavoidable factors, such as the depth of a drilled hole, changes the center of mass of the correction. The solution is to optimize the plane separation for the most critical unbalance level.

> **TIP** With variable unbalance amount it is often good to set the plane separation to be optimum for an unbalance 80% of the maximum value seen in production. The correction ratio is more critical with large corrections and the aim is to get as many parts in tolerance in one attempt.

In production – optimize the correction planes set in the measuring station with the actual correction cut (on hard bearing machine it does not affect calibration). Is the traverse set to give symmetrical movement with larger unbalance (plunge cut compared to plunge and travel)? On a short rotor does a maximum cut bring cutter or drill too close to other plane. Finally – is the correction repeatable?

## Is the process in control?

On a balancing machine the accuracy does not change with the amount of unbalance unless it is necessary to switch measuring ranges. If you try to measure the run to run variations as a percentage of the amount reading, or use the total angle variation, then the machine will appear to get worse when the unbalance is lower. This can be confusing.

In the theoretical case of zero unbalance the angle is completely random. If you are standing at the North Pole every direction is South! In practice the unbalance amounts may range from '0' to (say) 5% of tolerance but that is an almost infinite ratio.

Consider these two graphs. Which shows a better result? The graph on the left has an angle variation that is large and

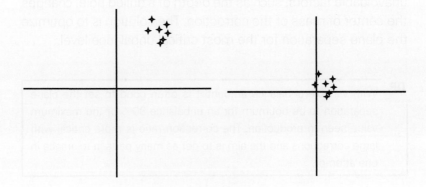

the percentage variation in the amount is also large. However also consider that all these points could be within tolerance! The graph on the right shows a much narrower range of angle and a smaller percentage variation of the amounts. It is possible that every one of these points could be out of tolerance.

Now for the problem – both graphs have the same cluster of data points. They were constructed with the same graphic file.

Statistical programs will give a machine a lower score if the data has unbalance that is too low. What we are really looking for is the machine accuracy at 1 × tolerance. If the unbalance is 5% of tolerance who cares if there is a 5% of tolerance error – that means the unbalance is between 0 and 10% of tolerance but this would also read as an error of 100% the average value.

A consistent magnitude of unbalance will give good readings to a statistical program since angle is ignored. A reading of 2.5 @ 90 is considered the same as 2.5 @ 270 degrees when in fact the change is 5.0. A second reading of '0' will show a change of 2.5 and score worse in a statistical program even though the vector change is half of that in the previous example.

---

**TIP** Adjust the unbalance of rotors used in testing so that it is close to tolerance and with actual numeric values closer to 99.9 than 100 (percentage error of one digit change is 10 times greater). If the tolerance is 1 gram and the rotors have 1.05 gram unbalance the precision is 0.01 gram. If we trim the unbalance to 0.985 the third digit represents a change of 0.001 gram. This can make a significant difference to a statistical analysis program.

## Process parameters

### *Gage R&R*

This is a standard test for measuring equipment. The procedure uses multiple tests on each of several parts with several
operators and qualifies the total errors as operator dependent
and machine dependant. If you are not familiar with gage R&R
then it is probably not significant for your operation (at the
present time).

Usually on a balancing machine there is very little operator variation. Due to the often extremely tight tolerances this test can
be a test of the rotor rather than the balancing machine. To
pass a gage R&R test at 10% of tolerance the machine must
be able to repeat to 2% of tolerance with the customer rotors.

---

**NOTE**

This is not a test of accuracy. A machine can pass this test and be 20%
off from true values. This test will show whether it is consistently off in
the same direction! This test is normally done after a calibration check
has verified that the results are correct.

---

A balancing machine is a precision measuring instrument. As
such it needs to be checked regularly for calibration, repeatability, etc. The Gage R&R test is not designed for a balancing
machine. It measures only the magnitude of the unbalance
with no reference to phase angle. If a rotor is measured and
the result is 0.5 g.mm at 120 degrees the unbalance is
0.5 g.mm. If it is re-measured and the result is 0.5 g.mm at
300 degrees the unbalance is 0.5 g.mm. For gage R&R pur-
poses the result is the same. Anyone familiar with balancing
understands that the vector difference between the two readings is 1.0 g.mm. Every test engineer I have dealt with on this
either never understood this or deliberately ignored it on the

basis that it was the specified test and had to be passed before the machine could ship. Preparation and selection of the test rotors used in the test may be a critical element in passing the test.

Gage R&R is checking repeatability. If the rotors have unbalance at 10× tolerance then the result has no connection to the machine's ability to determine whether a rotor is actually in tolerance. The rotors should be prepared so that the unbalance is close to the actual tolerance. If the tolerance is 1.00 then the rotors should be under this level since the display resolution will drop by one significant digit (the machine will read 1.01 or 0.999 and in one case the resolution is one part in a hundred while in the other it is one part in a thousand). This does not affect the machine accuracy but it does affect its ability to display the data.

Frequently, when doing a gage test, one rotor will show greater variation in readings than the others. Exchange the rotor for another or use one of the other rotors twice (measure it twice, not just copy the numbers). The test is to find the machine performance, not to find out how bad the rotors are. I have had machine buyoffs where I rejected all of the customer rotors. Often a customer will send reject rotors (for obvious reasons of economy) but the reason for rejection has to be unrelated to the balancing operation. If they are rejects for a bad thread on the end of the shaft that is OK, but rejects for bad bearing journals defeats the object of the testing. It is permissible to select the best ten rotors from a batch of (say) 50 because the test is to find out the limit of the machine with rotors that are as close to perfect as possible.

The test is following statistical methods. You don't have to know anything about statistics to do the test or calculate the results. If you run the first rotor twice and the results vary by more than

2% (of the tolerance) there is no point to continue since you will probably fail the test in another hour. Work on the root cause of the variation. Get the setup right and then do the test the result will be much less frustration and a machine that works better.

## *Cpk*

This is a test of process variability. It only applies to a balancer with a correction device since it measures the combination of measurement and correction. The big problem with this test related to balancing is that the incoming parts are not controlled. If one day the incoming rotors never exceed 5 × tolerance and the next day they average 15 × tolerance then suddenly the balancer doesn't work.

The problem is not the balancer but the initial unbalance. Actually the problem is in the definition of the process. If a balancer is rated for a correction ratio of 10:1 (ratio of initial to final unbalance) then the initial unbalance must be limited to 10 × tol. And anything higher than that must be excluded from the test.

With Cpk a batch of rotors with zero rejects is a different animal than a batch with one reject. No rejects and a maximum unbalance below 90% of tolerance and you are a hero.

If you are forced into using a Cpk test then the input condition must be controlled. First step is to sort the unbalanced rotors to eliminate any that are not correctable in one pass and those with very low unbalance. Keep the input data for the test. Note items such as shaft condition, varnish on laminations etc. Take pictures if necessary to document the test conditions.

It is important to remove the rotors with very low unbalance. These will have a result very close to zero unbalance which will show greater process variation. If you are sorting rotors the ideal is to find those that are about 5 times tolerance and widen this to between 3 × and 8 × tolerance. Every one of these should be in tolerance after balancing if the machine is working properly.

In balancing we are using a one-sided measurement. The minimum is zero and the maximum is tolerance. This covers some of the variations. For example, we have two rotors that have initial unbalance of 5 times tolerance. After correction they are both at 50% of tolerance. When we look closely at the data we see that one was corrected by 4.5 × tolerance and the other by 5.5 × tolerance. The balancing expert knows what the true machine performance actually is. The QA statistician sees two identical rotors since his system does not take into account the phase angle change due to balancing.

When you understand what is happening, you will be well aware that a Cpk number is meaningless on your balancing operation. Unfortunately, this does not help to get approval of more useful testing. By ensuring that you have a repeatable test condition you can verify the process at a later date.

Why so much fuss? The problem is not when you get a good score today but when you get a bad score next month! Cpk is a process monitor but you must define the process. This is critical!

The balancing process is not "measure and correct". A good process definition would be:

"Reduce initial unbalance of between 1 × tolerance and 5 × tolerance to within 97% of tolerance with no more than 5% of rotors exceeding 97% of tolerance."

If the input condition is controlled and you audit station does not reject more than 5% of parts you have achieved the process requirement. The use of 97% of tolerance means that on a rerun of the same rotors none that passed before should exceed tolerance.

---

**TIP**   If forced to do a Cpk test sort out twice as many rotors as you need for the test. Carefully pack half of the rotors so they will not degrade in storage and document this. When the machine fails a later test due to poor rotor conditions first check the machine very carefully then bring out the stored rotors and run them through the machine. It should come close to the original test and certify the machine while proving that the new rotors are the problem.

---

## Balancing efficiency

If we had the ability to buy perfectly homogenous material and machine it with perfect accuracy then there would be little need to balance rotors. The cost of such accuracy (if attainable) would be prohibitive.

By adding a balancing operation we can allow the material specifications, machining tolerances and assembly clearances to be relaxed and reduce the cost by a much greater amount than the cost of balancing. This is important to realize since many people consider balancing just to be an added cost rather than understanding it as a way to reduce overall product cost without reducing product quality. This can be carried too far. If the product arrives at the final balance station with too high a level of unbalance then the correction costs will be excessive, correction may damage the rotor or may affect its performance causing additional costs due to rejects.

Corrections such as drilled holes in a pump rotor may cause noise and/or cavitation, milling an armature may reduce performance, increase power losses and induce torque fluctu-ations (cogging). The problem may not be in the balancer but in the high initial unbalance.

Don't blame the balancer for the initial unbalance! I make a lot of fuss about this since I have had to visit customers and prove it countless times. I have heard the same story hun-dreds of times "The machine worked fine last week but today all we get is rejects." I shudder when I hear that the buyer got a good deal on bearings or rotor laminations. Even worse is to hear "A new supplier just came on line"

Don't try to correct the uncorrectable. If maximum material removal is 5 grams then don't even try to balance the rotor that has 6 grams of unbalance. The rotor is not 'scrap' until you make it so. The balancing machine can show when a pre-vious operation has a problem. Find the problem and fix it – the result will be lower initial unbalance. You can use the rotor with 6 grams of unbalance as evidence the scrap rotor is only evidence of murder!

## Balancing machine testing

### *How to be sure your machine does what you need.*
When purchasing a balancing machine the specification quoted will include the machine accuracy. This may be given is g.mm for a given rotor weight or in displacement of the mass center (say 0.1 micron) for a specified rotor weight range.

To test a balancer, a properly sized test rotor is needed. Normally the mass of the test rotor will be about 30% of the maximum weight capacity of the machine. Most machines do not run with rotors at the maximum weight capacity very often since purchasers like to have a safety margin in case they have to do a larger rotor in the future. More often machines are oversized and the average rotor is less than 30% of the machine capacity.

> **TIP** When buying a balancing machine consider how often you will need to balance the heaviest rotor(s). If you only have to do two a year you will be better to subcontract them and buy a smaller machine for the lighter rotors. Firstly, the capital investment is lower (pays for balancing several years production of the big rotor). Secondly, the smaller machine will be more sensitive, more accurate, quicker to setup (parts are smaller and lighter) and the minimum capacity will be lower (usually there are a lot more of the small parts to balance). The advantage here is that you will save time on every setup and have a more sensitive machine for those small rotors.

The test rotor enables the balancer to be compared to ISO or other standards and tests the actual capability of the machine under conditions that are close to ideal. An additional test with production parts is useful for "fitness for purpose" testing. The test rotor must have calibrated test weights of exact known mass.

A set of performance tests must be agreed. This is critical since the manufacturer may use a test procedure that does not necessarily prove that the machine is suitable for the specific rotor. The purchaser may be working from an unrealistic set of specifications due to a lack of knowledge of the real needs of the rotor – often taken from the specifications of a different rotor which may be larger, smaller or have different operating speed.

I am reminded of a turbocharger manufacturer who was unable to achieve assembly balance of a new model. We did a research project on the assembly characteristics and discovered that the clamping force holding the compressor rotor on the shaft was about 20% too high and this was distorting the shaft. No-one wanted to believe this of course and we were given the standard story about how they were the experts. Then someone noticed that the torque figure for the nut was taken directly from the specification of a larger size unit which had about 20% larger shaft diameter. Suddenly there was peace and quiet and the performance tests were passed easily.

## Balancer specifications

The first part of a set of specifications covers weight capacity, dimensional capacity and drive capacity. This is normally verified by visual inspection.

Performance testing uses a 'standard test rotor' and calibrated test weights. The mass of the test weights, location of mass center, and effective mounting radius must be verifiable and traceable back to National Standards. This is to establish the performance of the machine itself.

In order to meet published performance, or find that the machine does not meet the requirements, the testing has to

---

**NOTE**

A balancing machine measures the change of unbalance caused by adding a mass to the rotor. When the scales used to determine the test weight mass and the instrument used to measure the radius of its mass center are traceable then the paper trail is established and the machine calibration is therefore also traceable to the national standards.

be to a set of specific standards so that machines can be compared, verified and certified.

Before purchasing a machine the buyer should have determined the actual needs. After reading this book that should not be too hard!

The balancing machine manufacturer should verify that the data is complete before accepting the order. Unfortunately buyers often don't have complete specifications for what they want and machine manufacturers do not press for additional information in the desire to get the order logged. The problems do not arise until the machine is ready for testing and there are surprises. It is worth the effort to provide a final specification part to the machine manufacturer very soon after the order and communicate design changes promptly so the machine design can be updated. Conversely the machine manufacturer needs to document the initial agreement and the need for samples along with the results of the lack of representative samples.

**Test requirements and specifications**

When testing a balancing machine, or a process, we need to have a reference to guide us on what is a good result. The most widely used standard is ISO 1940 which provides an excellent guide to balancing tolerances based on just two parameters – type of application (turbine, grinder, diesel engine, electric motor) and maximum operating speed.

These two parameters can give us a balance tolerance referred to the displacement of the rotor mass center – eccentricity – in microns. $U = M \times e$.

Multiply eccentricity in microns by rotor mass in kg and result is the balance for that rotor tolerance in g.mm units.

Other industries, such as petrochemical, have their own specifications such as API 610 which uses the formula 4W/n which gives a result in ounce inches for rotor weight in pounds and speed in RPM.

There are a number of other specifications for specific industries – ARP 4048 for example in the aircraft industry – Aerospace Recommended Practices – Balancing Machines, horizontal, 2-plane, performance and evaluation. This is a complete balancing machine specification in itself. It gives machine sizes, drive capacity, accuracy, repeatability and rotor interface dimensions and an associated test procedure to verify performance.

For many applications there is no specific industry standard and a new machine is likely to be replacing one with significant age, less accuracy, and less features. In these situations the final consideration has to be "Is this reasonable?"

A balancing machine is a complex measuring instrument combined with mechanical drive components and rotor support components. It measures to accuracy equivalent to that of a high quality Coordinate Measuring Machine (CMM) and does so in shop conditions rather than those of a quality control laboratory.

## Balancing machine types

Balancing machines themselves are classified by four basic criteria:

### 1. Horizontal or vertical orientation of the rotor axis
Most horizontal balancers are used for rotors that have integral bearing journals that fit directly to the roller bearings of

the machine supports. After balancing a rotor can be removed from the machine, be replaced later and give virtually identical results.

Vertical balancers, and some horizontal machines, have a precision spindle as part of the machine. This provides a mounting location for rotor specific tooling. The rotors that get balanced on these machines do not have integral bearings and so the rotor must be mounted on some type of tooling.

Machines with built in spindles are normally limited in accuracy by the precision and repeatability of the tooling. This is important. A horizontal balancer used to balance a rotor with its own shaft and bearing journals can be balanced down to a mass eccentricity of 4–20 microinches (0.1–0.5 micron). This is the limit of the balancing machine to separate signal from noise. A rotor mounted on tooling can only be balanced to the repeatability of the tooling. It can be balanced to a low indication on the balancer but after removal and replacement will show a much higher level of unbalance.

Beware of assumptions on the real unbalance of a rotor when it becomes part of a system or assembly. After balancing the rotor may have rolling element bearings fitted. The inner bearing race runout will cause a change in the unbalance. The other problem with the rolling element bearing is that the results will be different with any change in the relative position of the inner and outer races. It is not uncommon for a rotor that is balanced to a high precision to exhibit excessive vibration in final test. When the bearings are removed and the rotor replaced in the balancer it shows to be good. Now bearings are refitted and the runout is likely to be at a different angle and the rotor may pass with flying colors. It is the bearings that cause the change and not the balancing machine.

### 2. Soft bearing or hard bearing

Soft bearing machines normally have a lighter construction and may have a lower purchase price but have to be calibrated for each rotor and each balancing speed. Hard bearing machines are permanently calibrated which makes for quicker setup and better performance with larger rotors, high drive powers and high windage loads.

Due to the heavier construction and increased need for a good foundation the initial cost of a hard bearing machine is usually higher than that of a soft bearing machine (especially with larger machine sizes). The hard bearing machine pays for itself with shorter setup times, lower balancing speeds (safety, drive power, reliability) and easier operation (less operator training).

### 3. Single or two plane

Static or dynamic – single plane balancers are usually vertical spindle machines. This is the ideal machine for disk shaped rotors with no shaft such as brake disks, flywheels and pump impellers. Most two plane machines have the capability to be set up for single plane balancing.

Narrow rotors only require single plane balancing. A narrow rotor is generally Ok for single plane balancing when is diameter is more than three times its width. We would also consider the distance between the bearings in service condition. For example a motor with bearings 21 apart may be fitted with a 6" diameter belt drive pulley. Even a wide pulley that we might consider two-plane balancing is narrow compared with the bearing spacing. If the pulley was 2.5 wide we may think it needed two-plane balancing but the application makes it logical to require only a single plane balance.

When we are dealing with high speed rotors such as those in small turbine engines the operating speed can be in the range of 50,000 rpm. Even very thin rotors can require two-plane balancing to very high tolerances. Related to this may be requirements for trim balancing assemblies of several rotors to ensure the engine performance is to specification.

### *4. Single plane balancers*
Further classified as rotating or non-rotating – centrifugal on gravitic (gravitational). Non-rotating machines are normally used for high volume production of rotors with coarse balance tolerances such as automobile brake drums – cycle time is short as no acceleration and deceleration and the tooling does not have to clamp the rotor just locate it.

The lack of centrifugal force to amplify the effect of the unbalance makes these machines suitable for rotors with relatively coarse tolerance. The additional requirement for short cycle times and simple tooling further makes the case for this type of machine in certain applications.

### Unbalance correction

Unbalance correction – there are only three possibilities – add, remove, or move mass.

Moving mass usually refers to weights in slots that can be re-positioned or unbalanced rings that can be rotated. This is good for high speed toolholders and grinding wheels that must be rebalanced often.

Material removal usually means drill, mill or grind. The problems for correction are possibility of weakening the rotor, or damaging the electrical properties. A drilled hole may cause windage or fluid noise.

Material addition can be a problem as well. Considerations such as weights coming off or rubbing on stationary parts of the assembly are involved. Adding a large mass to a weak structure can cause local bending.

If there is a large initial unbalance then correction is likely to be a problem. Few equipment designers are experienced in balancing and able to predict initial unbalance levels and properly design the appropriate provisions for correction into the rotor. The design of an allowance for correction material is in conflict with other requirements for minimum material, minimum size, minimum cost. Unfortunately it is usually when the design is complete, all the testing on prototypes is finished and the first production batch comes down the line that the truth is revealed – there is not enough material available to correct the unbalance. The other problem, related to the first, is that the initial unbalance is too high in relation to the tolerance for one pass correction.

Naturally the production models bear little relationship to the samples provided to the balancing machine manufacturer and also there were no production rotors available to test the machine before shipment.

It does not really matter what the cause really is – but it is vitally important to blame the balancing machine manufacturer for all deficiencies of the process and to refuse payment on the machine until they fix the problem! The problem cannot involve your employer until every other possibility has been exhausted!

Automated systems are custom tailored for the particular rotor, of family of rotors, to be balanced. Cycle time, method and amount of correction required, degree of automation and setup time are all part of the set of parameters that have to be optimized. The name of this game is investing capital upfront to save in unit costs over the life of the product.

If you need to obtain a new production balancing machine and if you want it to be delivered on time, have great uptime, and enhance your likelihood of promotion then just avoid the pitfalls above. The alternative is to get transferred to another project before the machine hits the floor, and other material hits the proverbial fan.

## In place balancing

One additional type of balance equipment is the portable balancer. Some situations require trim balancing in the final operating location. Large cooling fans require high drive power and are of large diameter which means that a large heavy duty balancing machine would be needed. In operation erosion and dirt build up change the unbalance so frequent rebalancing in the field is required to maintain performance.

Centrifugal pumps may have hydraulic unbalance – the mass of fluid in uneven passages changes the unbalance condition – this can only be fixed by trim balancing with measurements made under normal operating conditions.

High speed rotors may need to be trim balanced due to thermal and multiple bearing interactions. This is a subject for an advanced course.

A portable balancer comprises vibration pickup, zero phase reference and an instrumentation package. A calibration routine is used to calibrate the system for each specific rotor. Two-plane balancing is often practical but in some cases the rotor configuration does not lend itself well to consistent plane separation. A critical factor is that you have to determine that the problem really is unbalance and not looseness, misalignment or bad bearings.

Simple single plane balancing is not usually very difficult with a good piece of equipment available. Beware of the temptation to do two-plane trim balancing. Hire an expert first rather than after you have wasted much time and money trying to get it done with inexperienced staff.

If you have to use portable balancing equipment then the following will be useful.

## In place balancing procedures and pitfalls

A balancing machine is designed to isolate the unbalance signals. Precision support bearings eliminate bearing noise, the drive system is well balanced, and the work supports are designed to give good response to unbalance.

In place balancing is not so simple. The pickups typically measure a noisy signal and there may be high vibration signals with low unbalance. Noise can come from other machines, bad bearings loose mounting, worn drive shaft or damaged drive belt, gear noise and many other causes. In order to evaluate if balance is a problem the first step should be to check the total vibration. If this is low enough no action is required. If vibration is too high then adding a synchronous filter to the system will indicate what percentage of the vibration is at running speed. If the filtered signal is significantly lower than the total vibration then the problem is probably not unbalance.

If the filtered and unfiltered values are similar then unbalance is a probable cause. Attempting balance will soon reveal this. If the rotor does not respond to balancing attempts the next step is to check for other causes such as misalignment and maybe move the pickup from horizontal to vertical orientation to determine if it is a structural problem.

In short using portable balancing equipment is much more dependent on experience and diagnosis than shop balancing. Results are unpredictable. Keeping good records of each action and following careful procedures will help avoid frustration and loss of confidence and credibility later.

There are many applications where in-place balancing is the preferred method of balancing. Large pieces of equipment are difficult to move and the act of transport can distort them so that they are no longer in balance after assembly. These are not applications for the beginner! Usually this type of balancing is performed by highly trained and qualified people who have a wide experience of the machines being worked on and understand alignment, temperature effects, and results of load changes and the behavior of the bearing systems. These large machines often consist of a chain of several rotors coupled together and changing the unbalance of one unit can affect others in the chain.

In place balancing can be done using a simple portable system. Single plane balancing of a fan is relatively simple. Two-plane balancing of a large air handler in an industrial HVAC system is already getting complex. Vibration may be due to the motor, the fan or drive belts. It may be unbalance, misalignment, loose components or a damaged belt. At this stage we already need to analyze vibration and determine the frequencies where we have excessive vibration. The vibrations have to be traced to the various components and their relationships.

For large machines the balancing calculations are based on extensive computer analysis with a large database of historical information and test results from multiple correction planes and multiple speeds. Equipment manufacturers frequently collect data from each installation of a given machine type and use it to enhance the performance of future balancing, alignment, and other maintenance procedures.

Rotor assemblies with gearboxes have multiple speeds and factors such as coupling alignment. Bearing noise, gear noise and the equipment operation each produce characteristic vibrations. These have to be investigated and eliminated before balancing will be effective.

It is common for a specialist to be called in top balance a piece of equipment when the local maintenance team has given up. Often they are not sure whether they are doing something wrong or the whether the equipment is malfunctioning.

I have intense recollections of standing on top of a furnace in heavy snow checking a cooling fan that would not respond to balancing. We took two readings. One reading was of vibration at operating speed only (filtered by using a tachometer signal to synchronize) and the other was broadband (all frequencies). The total vibration was several times more than the once per rev. vibration. In addition the broadband vibration was reviewed using a FFT chart and showed frequencies related to bearing damage. Once the fan was fitted with new bearings the maintenance team had no trouble in balancing it. When we did a combination balancing operation and training course on the rebuilt fan we only had to deal with heavy rain. Plastic sheeting tents kept the equipment dry but did not help the balancing specialist much. I was careful to train the team well so they could do the other fans without assistance!

## Active balancing

Sometimes there is an unbalance problem that does not respond well to either the use of a balancing machine or to in place balancing.

One example of this would be high production grinding operations. The grinding wheels are large and fragile and run with continuous application of coolant. As the wheel wears a diamond tool is used to true and reshape the wheel. After a few hours of operation the wheel has developed unbalance. Stopping the wheel, removing it from the machine, balancing and re-assembly is not practical since there is the possibility of damage, there will be new runout from the assembly tolerances and the coolant drains to the bottom of the wheel making it impossible to obtain a correct unbalance reading.

The solution is to add the balancing operation to the grinding machine. This requires the addition of suitable vibration sensors, a balance correction ring that can be modified with the wheel running and a control system. In the 1980s a popular system used a system of water injection into four chambers to trim the unbalance. When the wheel was stopped the water drained out and the system was then ready for a new wheel. The main problems with this system were nozzles getting clogged and buildup of sediment in the chambers that limited unbalance correction. Conversely this was a very quick system that could rebalance a wheel in 5 seconds.

This method of correction was superseded by in the 1990s by closed systems that did not lose performance due to the build up of sediment. Two methods used with the closed system were movable mechanical masses and liquid (Freon or halon) that was moved from chamber to chamber by electrical heating. Both systems were less than perfect. Leakage of the mechanical systems led to sticking and lack of precision and the liquid system was slow to respond.

Now we are in the twenty-first century and the new generation of active balancing systems is fully sealed, fast responding and reliable – oh it is also able to work at speeds up to 42,000 rpm!

Today's active balancing systems are used on fans, turbines, pumps, machine tools and many other applications where maintaining accurate balance can eliminate costly premature shutdown or loss of performance.

A typical application is a process ventilating fan. If it stops, the line stops. As material builds up on the fan the unbalance gradually changes but the real problem occurs when a piece of the built up material breaks off – this causes a sudden large unbalance change that kills bearings and forces an immediate system shutdown costing many thousands of dollars.

By adding an active balancing system the gradual buildup of material can be compensated so bearings are not stressed. When material flies off the unbalance is automatically corrected in less time than it previously took to shut the system down. Savings are several times the cost of the balancing system every year!

Other applications include turbines that can be compensated for the effects of temperature and load change and can also be balanced during runup when they are close to a critical speed. By rebalancing during acceleration and deceleration the vibration levels can be maintained to safe levels even when passing through a resonance.

# 4

# Tooling and production

## Balance tooling

### Hydraulic expanding arbor

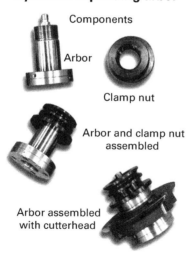

Components

Arbor

Clamp nut

Arbor and clamp nut
assembled

Arbor assembled
with cutterhead

Balance tooling is a necessary evil. It is the interface between the rotor and the balancing machine. Tooling may be as simple as a special set of bearings and even an end thrust support. Tooling may be as complex as a dummy engine rotor with safety/windage shroud and special drive adaptor.

Any device not supplied as standard equipment with the balancing machine is TOOLING. When you need a device to enable the rotor to be located in the machine for balancing you need TOOLING.

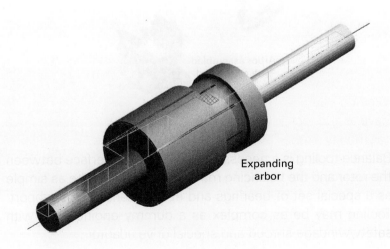

Expanding
arbor

## Like the care and feeding of monkeys

Don't you just love tooling? Looking after it is like looking after a barrel load of monkeys.

It is expensive – has to be light, accurate, repeatable, durable, special.
It adds weight – balancer sensitivity is related to rotor (and tooling) weight – this means that tooling reduces machine performance.
So far it costs a lot and reduces performance – what next?
Oh yes – it wears out, gets damaged and goes out of balance.

Why do you use it? Because you have no choice! When you have to add tooling to your balancer it is important to get value for money. A tool that looked cheap when you bought it can cause production delays, damage expensive rotors; cause test rejects – bad balance, and take too long to set up.

The point is that there are specific design features of good tooling that enable it to work efficiently and quickly and you need to know this. The following pages will help you with these decisions.

## Tooling solutions

The reason for using tooling is to locate a rotor in the balancing machine. It is critical that the requirements are considered in detail before selecting a given type of tool.

"... the balancing procedure can be as critical as the actual hardware used"

We have to consider the production volume, cycle time, and value of each rotor when evaluating tooling design. If the product we are balancing is an automotive brake disk the tolerance is coarse and the price of each rotor is low, but, there are thousands of rotors to balance. On the other hand we may be balancing the turbine of an aircraft engine where the volume is 50 parts per year. The fact that each rotor is valued at US$200,000, and that the cost of failure is a multiple of that, puts the tooling cost into perspective.

In the former case above the overriding concern is cycle time, but in the latter it is certification that the rotor is truly balanced to within the specification. From this we also realize that the balancing procedure can be as critical as the actual hardware used.

To evaluate a tooling design we must first look at the locating features available on the rotor. The requirement is to locate the rotor in a precise and repeatable way relative to the service center of rotation. To do this we need either two diameters or one diameter and a face.

Do not over-constrain the rotor or distortion/damage will result. If we attempt to use (say) two diameters and a face then

Turbine rotor located on two expanding diameter sections of a balance arbor.
Two diameters on the tool locate the arbor in the balancing machine.

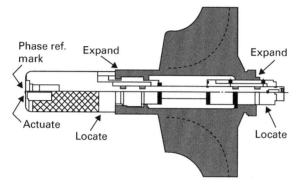

either the rotor or the tool will be deformed or stressed. The stress leads to both lack of accuracy and lack of repeatability since each rotor will have different dimensions and therefore a different stress level.

The tolerance on the rotor locating diameter(s) will control the tool design. If we are locating on a bore that has a tolerance of 0.005″ (0.127 mm) there is potential for errors of up to 50% of this amount if using a low cost solid arbor. If the tolerance permits the use of a solid arbor the cost of the tooling can be much lower.

In addition to the tolerance there is the locating and clamping method. If the rotor locates randomly on the arbor then repeatability becomes a major issue. If the clamp action pushes the rotor to one side consistently then we can have good repeatability even if there is an error due to the bore clearance.

---

**NOTE**

A known constant error can be compensated but a lack of repeatability cannot be compensated.

If the actual achieved locating tolerance multiplied by the rotor mass is greater than the unbalance limit then the tooling won't function to produce parts that are balanced to within the tolerance.

---

**NOTE**

A diameter with a tolerance of 0.01 mm will allow centering errors of max 0.005? eccentricity. If the tolerance would give 100% of 'U' then mounting error takes 50% of tolerance.

---

A good rule is to allow tooling to introduce errors no greater than 20% of tolerance. If this is not possible then the tolerance may not be valid. If the tolerance is valid then an alternate tooling design or balancing procedure may be required.

When assembly errors may exceed the balance tolerance consider the addition or substitution of a balance operation after the assembly operation.

To summarize the above points: the design of the tooling, manufacturing cost of the tooling, and maintenance/certification procedures are closely linked to both the rotor design and the tolerances on the location reference areas.


### Simple tooling

We always want to keep tooling simple, lightweight and low cost. For small batches of parts we can often use solid tooling. This often can be as simple as a shaft with an accurately machined diameter and locating face and a threaded section for clamping. The tooling accuracy is limited by the bore clearance of the rotor to the tooling. If the bore clearance is 0.02 mm then the accuracy is limited to 0.01 mm (half the clearance) since that is the maximum shift of the rotor from center. The repeatability can be twice that since the angle of the error is random.

## Balance tooling

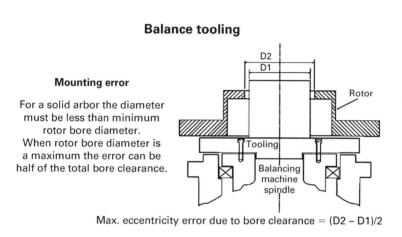

**Mounting error**

For a solid arbor the diameter must be less than minimum rotor bore diameter. When rotor bore diameter is a maximum the error can be half of the total bore clearance.

Max. eccentricity error due to bore clearance = (D2 – D1)/2

Solutions to the problem include making the tooling tapered (problem of keeping it perpendicular), having some expanding clamp action, or even designing a spring loading device to keep the offset to one side.

Where high accuracy is vital it is often possible to use a shrink fit procedure. This takes a long time but is practical with high value parts with low quantity. This type of tooling is used in the aircraft engine overhaul industry where the criteria of high value and low numbers certainly apply. The location diameters are often large compared to rotor weight, etc., and this would make expanding type tooling both heavy and expensive.

## Mechanical expanding tooling

For simple reliable production use, the mechanical arbor is hard to beat. It features simplicity, low cost wear parts and can be actuated in a number of different ways.

Typical layout is using a fixed cone with a moving ring or collet that clamps by means of axial travel. This has a secondary advantage that the clamping action pulls the rotor back against the axial location ring.

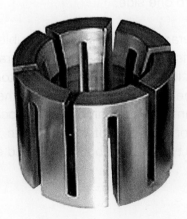

Expanding ring.

The moving ring is normally made of steel with slots to permit expansion. With careful design expansion as much as 1 mm in 50 mm can be achieved but normally expansion of about 0.25 mm.

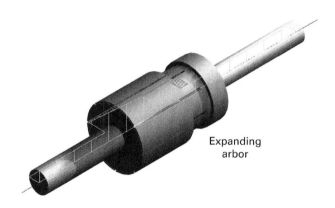

Expanding
arbor

Since the steel ring is radially incompressible the repeatability is related to the consistency and accuracy of the contact surfaces. Runout can be compensated. Designs can be made for internal or external clamping. By moulding rubber into the slots the clamping elements can be made resistant to penetration by chips and dirt. Regular maintenance (Cleaning and lubrication) is needed to prolong life of the system. For safety actuation is normally by means of springs – often disk springs – with pneumatic, hydraulic or mechanical release. The spring clamping means that no actuation connections are needed during the measuring cycle.

## Hydraulic expanding tooling

For very high accuracy the hydraulic expanding tooling works very well. The system uses grease inside a thin walled cylinder. Pressurizing the grease causes the walls of the cylinder to separate and conform to the arbor and the rotor.

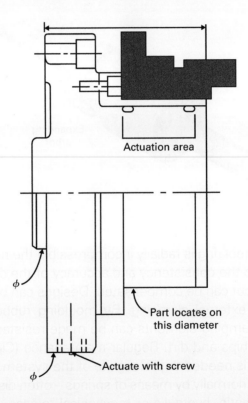

The compliance of the thin walls will compensate for slight imperfections in the rotor bore (scratches) that would cause an offset with a rigid tool. One restriction is that the expansion range is small (maybe 0.1 mm in 50 mm). Clamping is often done using a screw as a plunger but can be a spring-loaded plunger with powered release.

## Diaphragm type tooling

For large diameter parts a diaphragm type tool can be very effective. The diaphragm is like a pie plate with slits corresponding to the slices of the pie. Pushing up the center causes the outer rim to expand. The design can be made to expand or contract and also designed so that centrifugal forces from rotation increase the clamping pressure. Since this design is primarily for large diameters getting a clamping range of 0.2–0.3 mm is not difficult.

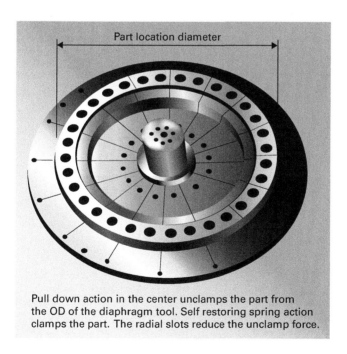

Part location diameter

Pull down action in the center unclamps the part from the OD of the diaphragm tool. Self restoring spring action clamps the part. The radial slots reduce the unclamp force.

## Production tooling

For disk brake rotors, tooling like that shown on page 193 is a classic application for a mechanical expanding tool.

For high volume production strong reliable tooling is needed.

In a machine with a vertical spindle, gravity helps locate the rotor against the axial face. A pull down action expands the tool while moving it downwards and securely seats the rotor. While a high clamping force enables quick acceleration and braking.

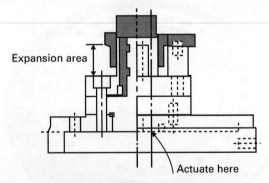

Tooling to locate a cam. Hydraulic expanding action is very accurate.

A wide expansion range enables the clamping to be disengaged for the correction cycle without having to remove the rotor from the measuring spindle. In a production environment the tolerance may not be that tight but the accuracy required from the tooling is still high. If gage R&R testing is required then the tooling and machine errors combined have to be less than 2% of the tolerance – often not easy!

# A closer look at tooling

Here we have a cross-section of the machine spindle. Note the housing and the spindle bearings, and finally the spindle itself. When the spindle is hollow there is provision to operate the tooling by using a pushrod or by just blowing air through the spindle to operate an air cylinder at the top. This illustration also appears on page 187, where we discussed a similar concept.

**Balance tooling**

**Mounting error**

For a solid arbor the diameter must be less than minimum rotor bore diameter. When rotor bore diameter is a maximum the error can be half of the total bore clearance.

D2
D1

Rotor

Tooling

Balancing machine spindle

Max. eccentricity error due to bore clearance = (D2 – D1)/2

The rotor is located by one diameter and the back face of the mounting flange (not the lower surface since the disk is located on the flange in service). In this case we have a solid tool with no provision for clamping by expansion. Locating error is the same as radial clearance (half of the diameter difference).

With a tool such as this there is a large random error that cannot be compensated and therefore the minimum achievable unbalance of the actual rotor can be more than the tolerance. To make the tooling more usable one must reduce the random error possibilities.

Providing a spring preload to one side will reduce the random error but add a constant error. A constant error can be compensated and the net result is a lower achievable unbalance of the rotor.

---

**NOTE**

We do not achieve lower readout numbers on the balancing machine. The difference here is that when a balanced rotor is rechecked (removed from the tooling and replaced in a random position) the (average) unbalance will be lower.

When we qualify tooling the procedure is simple. The machine is loaded with a rotor which is balanced to just within tolerance. The machine is run at least five times without touching the rotor. The results give us the machine repeatability. The machine must be capable of repeating to less than 10% of tolerance! Now the rotor is removed and replaced and the machine run. This is repeated at least fives times. The variation in the results is the total variation. We can deduct the machine variation to see the tooling variation.

---

When the tooling is not performing as required things become less simple!

## Expanding tooling (production application)

Look at the expanding type tool opposite. The central part is the measuring station tooling while the outer jaws hold the disk for correction by milling. With the measure tooling unclamped milling stress is not transferred to the measuring spindle or tooling. The three jaw chuck holds the rotor securely for milling and is connected to an indexer that allows the rotation needed for unbalance correction.

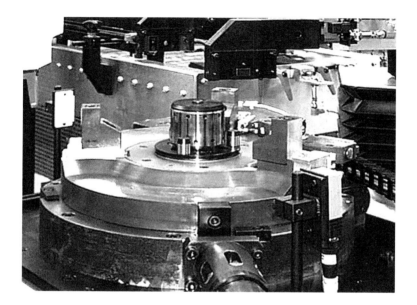

The measure tooling expands by means of die springs built into the tooling package. It is released by a pushrod through the center of the spindle. This is fail safe and avoids the need to have any connection to the spindle during measuring. As there might be stray chips from the milling operation the expanded collet is molded with a soft oil-proof rubber compound to ensure that chips do not get trapped in the expansion slots.

---

**NOTE**

This type of tooling gives much more precise results than a solid arbor. It also has very short clamp and unclamp times. This tooling costs more! It is the old dance between cost, cycle time and performance.

---

# Selection of tooling

There is a vital difference between runout and repeatability. The runout is a constant and therefore can be compensated. If it is too high the resulting unbalance can overload the measuring system so the tool needs to be centered accurately (approximately 0.0004″ or 0.01 mm).

Inconsistency is the enemy of good balance. Tooling design has to have the emphasis on repeatability. Ensuring that length to diameter and diameter of locating cone to bore diameter are optimized are primary design factors. The actuating mechanism must be tightly controlled. It is not enough that the rotor is located accurately. The tooling package in its entirety must be located to close limits or the results from run to run will be different.

Many seemingly small items have to be optimized to arrive at a useful, accurate, repeatable tool package. Some of the design features of the expanding tool are mentioned here.

To actuate (unclamp) the tool, compressed air is introduced inside the machine's hollow spindle. This avoids the mass of pushrods, etc., but the cylinder on top of the spindle has to be light, compact and stable and must also generate high actuating forces. To achieve this we have a short stroke and large diameter. Friction must be minimized and the clamping springs must be matched so the piston stays horizontal. We also use air to unclamp when the spindle is stopped and springs for failsafe clamping.

### Example calculations

Look back at the example of balance tooling. For a rigid tool the error is random with errors up to 80% of tolerance. Unless the exact diameter of the bore is known the maximum error is unknown. The angle of the error is also unknown. To ensure that the unbalance is within tolerance after a recheck – the reading on the machine after the rotor is removed and randomly replaced – rotor must be balanced to less than 20% of tolerance.

An expanding arbor eliminates the bore clearance. The error is now only the repeatability in clamping which is 8% of tolerance. If the rotor is balanced to less than 82% of tolerance it will repeat within tolerance after removal and replacement.

**NOTE**
For simplicity, it is assumed that the machine itself repeats perfectly. In practice, an additional allowance has to be made for the balancing machine run to run variation. If this was 10% of tolerance the rigid tooling would then require that the rotor be balanced to 10% of tolerance!

The point here is that the extra cost of the expanding arbor will be reclaimed by the reduction in time to balance a certain number of rotors. This will be the payback time for the investment.

# Runout compensation

We have mentioned that runout errors can be compensated. Here is where we show you how.

**IF** the repeatability is OK (sometimes a big IF) the constant error can be compensated either mechanically or electrically.

In either case what we are doing is separating the unbalance related to the spindle and/or tooling from the unbalance of the rotor and compensating the spindle and tooling runout and unbalance.

Making a mechanical correction to the tooling to compensate for runout is more time-consuming than doing it in the instrumentation but is not subject to errors due to drift or recall of wrong data. For tooling, such as balance arbors, the mechanical compensation is referred to as 'biasing'. A biased tool run without a rotor will probably read as out of balance – that is because the compensation is for the runout of the mounting locations as well as the unbalance.

## Graphic representation – 1

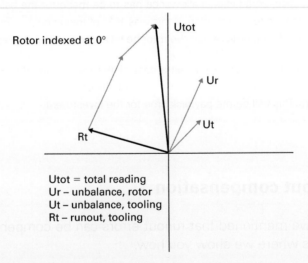

Utot = total reading
Ur – unbalance, rotor
Ut – unbalance, tooling
Rt – runout, tooling

Here we see the reading of a rotor mounted on tooling (single plane situation). The total unbalance reading is comprised of the rotor unbalance plus the tooling unbalance and the unbalance resulting from tooling runout.

If the machine was run without the rotor the only reading would be the tooling unbalance, which is the smallest of the three amounts. By plotting the polygon of the three amounts we obtain the total. Of course the only two data points we can get are Ut and Utot.

In order to isolate the unbalance of the rotor we need to rotate it on the tooling by a known amount – 180 degrees is the standard default angle but with microprocessor instrumentation other angles can be used.

## Graphic representation – 2

Now we have the rotor re-oriented. Utot is the new reading made up from Ur, Ut and Rt, and we can ask ourselves "what is different from previous readings?"

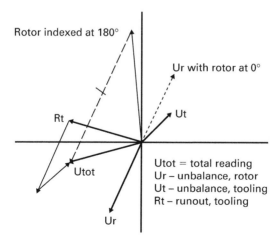

The rotor unbalance has changed phase by 180 degrees! The difference in the readings is exactly two times the unbalance of the rotor. If doing a manual correction, we would have balanced the readings to 'zero' at the first stage. At this second stage the unbalance reading would be two times the rotor unbalance. By making a correction to 50% of the value, and holding the phase the same, the actual rotor unbalance would be corrected. The other 50% of the correction is then made on the tooling to bring the reading to 'zero'. At this point the rotor and the tooling should both be balanced. Rotating the rotor on the tooling to any random angle should still give a machine reading of 'zero'.

### A quick review

The rotor installs repeatably to the tooling. Adding the rotor does not change the unbalance of the tooling. Runout of the tooling causes a 'phantom' unbalance to be displayed on the balancer. When we rotate the rotor on the tooling the difference between reading 1 and reading 2 is that the heavy spot on the rotor has moved 180 degrees. By adding a mechanical (or electronic) correction to the readings we can eliminate the effect of the runout.

When we turn the rotor 180 degrees the tool unbalance does not change, and the phantom unbalance from the tooling runout does not change. When this becomes clear in the mind (it does take a while sometimes) then compensation is easy.

---

**NOTE**

There is a big potential problem with electronic compensation. It relies on the phase reference being absolutely stable. On a vertical spindle machine, or an end drive machine where a magnetic sensor picks up a keyway or drilled hole, this is normally very stable. On a horizontal machine the photocell can get bumped and then the compensation is no longer valid. In a production situation it is best to compensate 90% of the error mechanically to minimize this problem.

---

## Other types of tooling

Some tooling does not have the runout problem – support tooling that enables the rotor to be supported on its own bearings.

A rotor like the APU shown above can be balanced more accurately in its own bearings because the errors of runout of the inner race are eliminated and the bearings are preloaded to operational condition. The tooling has to be adapted to the balancing machine to preserve two-plane measuring with close plane separation. In this case the rotor is air driven and the bearings are pre-lubricated with turbine engine oil with balancing taking place at a speed in excess of 2,000 rpm. With the hard bearing technology used in this example the exact speed is not important since the machine setup is not affected by speed changes provided the speed is held constant during the 5 seconds or so of actual unbalance measurement.

## More tooling

In the example below the rotor runs on rollers and could be balanced on a standard balancing machine except for the fact that the rotor has to be assembled with the non-rotating stator before final balancing. The tooling in this case has to locate the stator precisely with respect to the rotor in radial and axial directions. The belt drive is accommodated by a slot in the tooling (the belt is applied over the rotor before it is loaded to the tooling).

To use the tooling design in this or the previous illustration the machine has to be equipped with a support system that allows torsional freedom of the tooling mount. Without this the machine would not be able to separate signals such as bearing swash (axial bearing runout) from unbalance (bearing error on the left side causes a synchronous force at the right support). This phantom unbalance is proportional to speed but real unbalance varies with speed squared on a hard bearing machine – the phantom unbalance changes with speed so the machine gives different readings at different speeds.

The other effect is a loss of plane separation since the rigid connection causes the two supports to see the same signal. The supports with torsional freedom are usually configured with upper sections having cylindrical bores to accept matching tooling rings which in turn locate the engine bearing OD. The rotors illustrated here have similar balancing configurations. The rotor rolling element bearings are located in a rigid frame which keeps them properly aligned.

The tooling is mounted to the balancer supports with a device that allows the work support to freely rotate through a small angle and comply with the mounting dimensions of the tooling. With this configuration the machine reacts as if the rotor were supported on rollers at the supports even if the rotor bearings are (say) at one end and the center. The dimensions for plane setting are based on the distance from the center of the work supports to the correction planes and the positions of the actual rotor bearings.

# Machine tool cutter balancing

### Vertical spindle balancer

A vertical spindle balancing machine is configured in the exact same way as with the previous horizontal machine. The spindle is supported in bearings, which in turn are located in a suspension that includes the pickups. This particular machine is for balancing toolholders but the only difference here is the taper bore in the spindle.

In all cases of vertical balancers, minimum spindle runout (radial and axial) is critical to enable precise location of the tooling. The bearings are designed around a given weight capacity. The configuration and accuracy of the bearings determine the machine accuracy and repeatability and are typically considerably more precise than the tooling repeatability except at the lower weight limit of the balancer (this is a constant while the unbalance tolerance varies with rotor weight).

Since we are dealing with toolholders in this example, the illustration opposite shows the situation for a tool and toolholder. The rotational axis is defined by the upper and lower sections of the taper in the spindle.

The tooling (spindle) and its bearings define the rotational axis – or bearing axis. In this situation we cannot use or determine

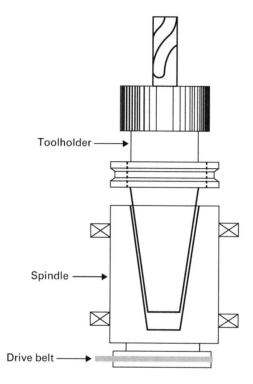

the actual bearing axis of the rotor. The bearing axis is defined by the tooling. This applies to any tooling configuration. Sometimes the location is defined by two diameters (as in this case) and on others, such as the brake disk, it is defined by a diameter and a face. With this type of toolholder there is only the need for single-plane balancing since the toolholder is small and short in comparison to the spindle that it mounts into.

To correct unbalance we need to change the mass distribution and that means adding mass – such as screws, washers, using a longer screw, or welding a weight in place. On a piece of tooling, balancing will be needed multiple times over its lifetime and balance correction should not damage the tool.

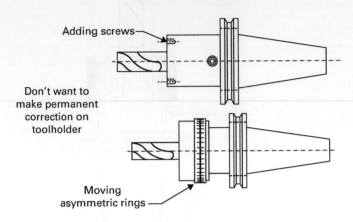

**Correct unbalance by:**

Adding screws

Don't want to make permanent correction on toolholder

Moving asymmetric rings

Balanceable Toolholders usually have tapped holes axially or radially, or have a pair of rings that can be rotated to give varying amounts of unbalance. Typically, the rings are predrilled with about 10 holes over a 60 degree arc and pre aligned so they cancel each other out. Rotating the ring against each other, from a few degrees to 180 degrees, adjusts the amount. Adjusting both together changes the angle. The rings are locked with setscrews.

With special designs such as long boring bars two-plane balancing is required but the rotor mass and correction areas are normally close together. The main part of the toolholder is solid, accurately machined and stable and the unbalance normally comes from the clamping mechanism and the tool. It does make sense to apply the correction close to the location of the unbalance.

## HSK tooling

Ever increasing spindle speeds, the requirements for greater rigidity, and certain specific problems have led to a recent explosion in new types of interface to mount the cutting tool to the spindle. The European development of the HSK system is becoming the most popular and widely accepted of new designs.

Some of the specific problems with standard taper mount toolholders (also called steep taper, MT, CAT) are directly related to high speed operation and high cutting forces. Since there is a long taper length it is impossible to have any spindle match with any toolholder and contact along the entire length. If first contact is made at the small end of the taper the tool will wobble. If contact is made first at the large end of the taper the spindle will be distorted, and the bearing inner race expanded leading to hot bearings. High speed bearings have precise clearances and preloads and can easily be damaged by this condition.

With the super accuracy of today's machines another problem arises. Once the spindle has been to high speed, the centrifugal force on the spindle tends to increase the spindle bore diameter and the toolholder moves further into the spindle. Now the preset dimensions have changed. A logical other problem is that it is now harder to remove the toolholder from the spindle.

# The HSK brings new factors into workholding

The HSK toolholder is made with a thin wall so it can collapse rather than force the spindle to expand. The HSK toolholder is pulled back against a face which means it cannot move further back into the spindle and cannot 'wobble' and become eccentric. With high clamping force against the face there is greater torque capacity. For low speed high torque application (where balancing is not critical anyway) a positive drive key is used.

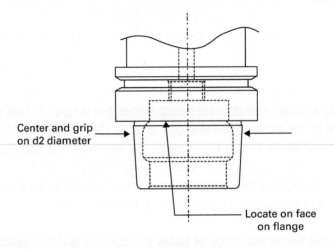

Center and grip on d2 diameter

Locate on face on flange

The main benefits are:

High speed does not change tool position. Conventional tapers expand at high speed and tool moves back into taper – hard to remove and changes tool reference!

More accurate location – using the face keeps the tool square. This is especially important with longer tools.

HSK allows higher spindle speeds – and therefore can require tighter balancing tolerances.

HSK, however, is not a magic bullet. Some other details like a higher clamping force can increase torque transmission. The downside to this is the more complex clamping system (drawbar assembly) which can cause unbalance problems at the upper end of the spindle. Sizes are supposed to be standard but many companies have their own versions.

HSK is combined with other technologies. There is another interface that affects balancing – that between the toolholder and the cutting tool. Together with HSK technology manufacturers have developed low mass and high accuracy interfaces using lobed tools, heat and shrink clamping, hydraulic unclamping.

For safety and accuracy we normally use a vertical balancing machine with an HSK spindle insert and a power drawbar. The machine accuracy is normally limited by the repeatability of tool location and drawbar actuation rather than be the measuring precision of the balancer.

With high speed machining balancing goes in and out of popularity. As speeds increase balancing becomes more critical but then tool mounting technology improves to eliminate the need for 're-balancing'. Then speeds increase again.

This is how many industries and balancing applications function in the long term. Balancing is perceived as an added cost and designed out of the production sequence. Then there is a need to improve performance or reduce mass and suddenly balancing is needed (usually when the new design does not perform as expected).

# INDEX

Note: The locators of entries taken from floats are set in *italics* font.

Printed and bound by CPI Group (UK) Ltd, Croydon, CR0 4YY

08/05/2025

01864850-0003